LECTURES ON DIFFERENTIAL AND INTEGRAL EQUATIONS

KÔSAKU YOSIDA
University of Tokyo

DOVER PUBLICATIONS, INC.
New York

This Dover edition, first published in 1991, is an unabridged and unaltered republication of the work first published by Interscience Publishers, New York, in 1960.

Manufactured in the United States of America
Dover Publications, Inc., 31 East 2nd Street, Mineola, N.Y. 11501

Library of Congress Cataloging-in-Publication Data

Yosida, Kôsaku, 1909–
    [Sekibun hoteishiki ron. English]
    Lecturas on differential and integral equations / Kôsaku Yosida.
            p.   cm. — (Pure and applied mathematics ; v. 10)
    Translation of: Sekibun hoteishiki ron.
    Reprint. Originally published: New York : Interscience Publishers, 1960.
    Includes bibliographical references and index.
    ISBN 0-486-66679-4
    1. Differential equations. 2. Integral equations. I. Title. II. Series: Pure and applied mathematics (Interscience Publishers) ; v. 10.
QA372.Y613   1991
515'.35—dc20
90-28876
CIP

Foreword

The present book is the English edition of my book published originally, in Japanese, by the Iwanami Shoten in the series "Iwanami Zensho." It was intended to be a self-contained exposition of the theory of ordinary differential equations and integral equations. It especially gives a fairly detailed treatment of the boundary value problem of second order linear ordinary differential equations, and includes an elementary exposition of the theory of Weyl-Stone's eigenfunction expansions in the form completed by Titchmarsh-Kodaira's formula concerning the density matrix of the expansion.

The author wishes to express his sincere thanks to Professor Shigeharu Harada of Chiba Institute of Technology for his time and effort in preparing the English translation of this book, and to Professor Lipman Bers of New York University for his kind suggestion to include this book in his series.

The author also extends his cordial thanks to the Iwanami Shoten and the Interscience Publishers, Inc., who kindly agreed to the publication of the English edition, and to the International Academic Printing Company for the painstaking work of printing the book.

August, 1960   KÔSAKU YOSIDA

Contents

CHAPTER 1

THE INITIAL VALUE PROBLEM FOR ORDINARY DIFFERENTIAL EQUATIONS

An *nth order ordinary differential equation* is a functional relation

$$F\left(x, y, \frac{dy}{dx}, \frac{d^2y}{dx^2}, \cdots, \frac{d^ny}{dx^n}\right) = 0$$

between a variable x, an unknown function y, and its derivatives

$$\frac{dy}{dx}, \frac{d^2y}{dx^2}, \cdots, \frac{d^ny}{dx^n}$$

A function $y(x)$ which satisfies this relation is called a solution of the differential equation. For example, for any constant C, the function

$$y = \sin(x - C)$$

is a solution of the first order equation

$$\left(\frac{dy}{dx}\right)^2 + y^2 - 1 = 0$$

In general, the *general solution* of an *n*th order equation contains *n* arbitrary constants $C_1, C_2, \cdots, C_n$. These constants may be determined by so-called *initial conditions* at a point $x = x_0$, which are the prescribed values of $y, dy/dx, \cdots, d^{n-1}y/dx^{n-1}$ at the point $x = x_0$. The solution thus obtained is called a *particular solution*.

In this chapter, we shall show that the *initial value problem* for differential equations can be reduced to a system of integral equations and solved by the method of successive approximations. This chapter is intended as a prerequisite for the theory of the differential equations as considered in the following chapters.

§1. Successive approximations

1. *Existence and uniqueness of the solution of the ordinary differential equation of the first order*

An ordinary differential equation of the first order is generally written in the form

(1.1) $$F(x, y, dy/dx) = 0$$

In the following, we shall restrict ourselves to those cases in which (1.1) can be solved for dy/dx and written in the form

$$(1.2) \qquad\qquad dy/dx = f(x, y)$$

The simplest case of (1.2) is of the form

$$(1.3) \qquad\qquad dy/dx = f(x)$$

The solution of (1.3) is given by

$$(1.4) \qquad\qquad y(x) = \int_{x_0}^{x} f(t)\,dt + C$$

in the region where $f(x)$ is continuous. The integration constant C is determined by the value of $y(x)$ at $x = x_0$, that is,

$$(1.5) \qquad\qquad y_0 = y(x_0) = C$$

Accordingly, the solution of (1.3) satisfying the condition $y = y_0$ at $x = x_0$ is given by

$$(1.6) \qquad\qquad y(x) = y_0 + \int_{x_0}^{x} f(t)\,dt$$

The condition,

$$(1.5') \qquad\qquad y = y_0 \quad \text{at} \quad x = x_0$$

is called the *initial condition* for the solution of the differential equation (1.3).

Our purpose is to find a solution of the general equation (1.2) subject to the initial condition (1.5'). To formulate the problem precisely, we make the following assumptions concerning $f(x, y)$.

ASSUMPTION 1. The function $f(x, y)$ is real-valued and continuous on a *domain*[1] D of the (x, y)-plane given by

$$(1.7) \qquad x_0 - a \leqq x \leqq x_0 + a, \qquad y_0 - b \leqq y \leqq y_0 + b$$

where a, b are positive numbers.

ASSUMPTION 2. $f(x, y)$ satisfies the *Lipschitz condition* with respect to y in D, that is, there exists a positive constant K such that

$$(1.8) \qquad\qquad |f(x, y_1) - f(x, y_2)| \leqq K|y_1 - y_2|$$

for every pair of points $(x, y_1), (x, y_2)$ of D.

---

[1] By a domain, we mean a nonempty open connected set, and by a closed domain, the closure of a domain. However, we shall use the word "domain" to denote "closed domain," when the domain is explicitly defined as (1.7).

Remark. The second assumption would seem to be less natural than the first, but its importance will become clearer in the following proof.

It is worth while to show that if $f(x, y)$ has a continuous partial derivative $\partial f(x, y)/\partial y$ on D, then $f(x, y)$ satisfies the second assumption. This is seen as follows. Since $|\partial f(x, y)/\partial y|$ is continuous on the bounded closed domain D of the (x, y)-plane, $|\partial f(x, y)/\partial y|$ is bounded on D. Put

$$(1.9) \qquad K = \sup_{(x,y)\in D}\left|\frac{\partial f(x, y)}{\partial y}\right|$$

Then the mean-value theorem implies that (1.8) holds for $f(x, y)$.

By Assumption 1, $|f(x, y)|$ is continuous on the bounded closed domain D, therefore $|f(x, y)|$ is bounded on D, that is,

$$(1.10) \qquad \sup_{(x,y)\in D}|f(x, y)| = M < \infty$$

Set

$$(1.11) \qquad h = \min(a, b/M)$$

Let us define a sequence of functions $\{y_n(x)\}$ for $|x - x_0| \leqq h$, successively, by

$$(1.12) \qquad \begin{aligned} y_0(x) &= y_0 \\ y_1(x) &= y_0 + \int_{x_0}^{x} f(t, y_0(t))\,dt \\ y_2(x) &= y_0 + \int_{x_0}^{x} f(t, y_1(t))\,dt \\ &\cdots\cdots\cdots\cdots\cdots\cdots\cdots \\ y_n(x) &= y_0 + \int_{x_0}^{x} f(t, y_{n-1}(t))\,dt \end{aligned}$$

The theorem to be proved is that $\{y_n(x)\}$ converges uniformly on the interval $|x - x_0| \leqq h$, and the limit $y(x)$ of the sequence is a solution of (1.2) which satisfies (1.5′).

Proof. According to (1.10) and (1.11), we obtain

$$|y_1(x) - y_0| \leqq hM \leqq b$$

for $|x - x_0| \leqq h$. Therefore $\int_{x_0}^{x} f(t, y_1(t))\,dt$ can be defined for $|x - x_0| \leqq h$, and

$$|y_2(x) - y_0| \leqq hM \leqq b$$

In the same manner, we can define $y_3(x), \cdots, y_n(x)$ for $|x - x_0| \leq h$, and obtain

$$|y_k(x) - y_0| \leq hM \leq b , \qquad k = 1, 2, \cdots, n$$

Using Assumption 2, we have

$$|y_{k+1}(x) - y_k(x)| \leq K \left| \int_{x_0}^{x} |y_k(t) - y_{k-1}(t)| \, dt \right|$$

for $|x - x_0| \leq h$. Therefore, if we assume that, for $k = 1, 2, \cdots, n$,

$$(1.13) \qquad |y_k(x) - y_{k-1}(x)| \leq \frac{b(K|x - x_0|)^{k-1}}{(k-1)!} \qquad \text{for } |x - x_0| \leq h$$

we obtain, for $k = n + 1$,

$$(1.14) \qquad |y_{n+1}(x) - y_n(x)| \leq \frac{b(K|x - x_0|)^n}{n!} \qquad \text{for } |x - x_0| \leq h$$

Since (1.13) holds for $k = 1$ as mentioned above, we see, by mathematical induction, that (1.14) holds for every n. Thus, for $m > n$ we obtain

$$(1.15) \qquad |y_m(x) - y_n(x)| \leq \sum_{k=n}^{m-1} |y_{k+1}(x) - y_k(x)| \leq b \sum_{k=n}^{m-1} \frac{(Kh)^k}{k!}$$

Since the right side of (1.15) tends to zero as $n \to \infty$, $\{y_n(x)\}$ converges uniformly to a function $y(x)$ on the interval $|x - x_0| \leq h$. Since the convergence is uniform, $y(x)$ is continuous and moreover, evidently, $y(x_0) = y_0$. To prove that $y(x)$ is a solution, we use the following known fact: If the sequence of functions $\{y_n(x)\}$ converges uniformly and $y_n(x)$ is continuous on the interval $|x - x_0| \leq h$, then

$$\lim_{n \to \infty} \int_{x_0}^{x} y_n(x) \, dx = \int_{x_0}^{x} \{\lim_{n \to \infty} y_n(x)\} \, dx$$

Hence we obtain

$$\begin{aligned}
y(x) &= \lim_{n \to \infty} y_{n+1}(x) \\
&= y_0 + \lim_{n \to \infty} \int_{x_0}^{x} f(t, y_n(t)) \, dt \\
&= y_0 + \int_{x_0}^{x} \{\lim_{n \to \infty} f(t, y_n(t))\} \, dt \\
&= y_0 + \int_{x_0}^{x} f(t, y(t)) \, dt
\end{aligned}$$

that is,

$$(1.16) \qquad y(x) = y_0 + \int_{x_0}^{x} f(t, y(t))\, dt\,, \qquad |x - x_0| \leqq h$$

The integrand $f(t, y(t))$ on the right side of (1.16) is a continuous function, hence $y(x)$ is differentiable with respect to x, and its derivative is equal to $f(x, y(x))$, q.e.d.

The method of the above proof is called the *method of successive approximations*, or *Picard's method*.

Integrating from x_0 to x, we see that a solution $y(x)$ of (1.2) satisfying the initial condition (1.5') must satisfy the *integral equation* (1.16). The above proof shows that this integral equation can be solved by the method of successive approximations.

Uniqueness of the solution. We have obtained, by the method of successive approximations, a solution $y(x)$ of (1.2) satisfying the initial condition (1.5'). However, there remains another important problem, the problem of uniqueness: Is there any other solution satisfying the same initial condition? If the solution is not unique and there are solutions other than the $y(x)$ obtained above, we must find other methods to obtain them. Fortunately, under our two assumptions, we can prove the uniqueness of the solution. To see this let $z(x)$ be another solution of (1.2) such that $z(x_0) = y_0$. Then

$$z(x) = y_0 + \int_{x_0}^{x} f(t, z(t))\, dt$$

By Assumption 2, we obtain

$$(1.17) \qquad |y(x) - z(x)| \leqq K \left| \int_{x_0}^{x} |y(t) - z(t)|\, dt \right|$$

Therefore we also obtain for $|x - x_0| \leqq h$

$$|y(x) - z(x)| \leqq KN |x - x_0|$$

where

$$N = \sup_{|x-x_0| \leqq h} |y(x) - z(x)|$$

Substituting the above estimate for $|y(t) - z(t)|$ on the right side of (1.17), we obtain further

$$|y(x) - z(x)| \leqq N(K|x - x_0|)^2/2!$$

for $|x - x_0| \leqq h$. Substituting this estimate for $|y(t) - z(t)|$ once more on the right side of (1.17), we have

$$|y(x) - z(x)| \leqq N(K|x - x_0|)^3/3!$$

for $|x - x_0| \leqq h$. Repeating this substitution, we obtain

$$|y(x) - z(x)| \leqq N(K|x - x_0|)^m/m! \qquad m = 1, 2, \cdots$$

for $|x - x_0| \leqq h$. The right side of the above inequality tends to zero as $m \to \infty$. This means that

$$N = \sup_{|x-x_0| \leqq h} |y(x) - z(x)|$$

is equal to zero, q.e.d.

EXAMPLE. To illustrate the general procedure, we shall solve the differential equation

$$dy/dx = y$$

under the initial condition $y(0) = 1$ by the method of successive approximations. We have from (1.12)

$$y_1(x) = 1 + \int_0^x dt = 1 + x$$

$$y_2(x) = 1 + \int_0^x (1 + t)\, dt = 1 + x + \frac{x^2}{2!}$$

$$\dots \dots \dots \dots \dots \dots$$

$$y_n(x) = 1 + \int_0^x \left(1 + t + \frac{t^2}{2!} + \cdots + \frac{t^{n-1}}{(n-1)!}\right) dt$$

$$= 1 + x + \frac{x^2}{2!} + \cdots + \frac{x^n}{n!}$$

In this way we obtain the well-known formula

$$y(x) = \lim_{n \to \infty} y_n(x) = \exp x = \sum_{n=0}^{\infty} \frac{x^n}{n!}$$

2.　*Remark on approximate solutions*

Letting $m \to \infty$ in (1.15), we obtain

$$(2.1) \qquad |y(x) - y_n(x)| \leqq b \sum_{k=n}^{\infty} \frac{(Kh)^k}{k!}$$

for $|x - x_0| \leqq h$. The equation (2.1) is an estimate of the error of the nth approximate solution $y_n(x)$. In view of (2.1), the method of successive approximations may be used, in principle, to obtain an approximate solution to any degree of accuracy. This method, however, is not always practical because it requires one to repeat the evaluation of indefinite integrals many times.

We shall now consider another method which is sometimes rather useful. Suppose that $g(x, y)$ is a suitable approximation to $f(x, y)$

such that we can find the solution $z(x)$ of the differential equation

(2.2) $$dz/dx = g(x, y)$$

on the interval $|x - x_0| \leq h$ satisfying the initial condition $z(x_0) = y_0$.
We put

(2.3) $$\sup_{(x,y) \in D} |f(x, y) - g(x, y)| \leq \varepsilon$$

Let $y(x)$ be the unique solution of the differential equation

(2.4) $$dy/dx = f(x, y)$$

on the interval $|x - x_0| \leq h$ satisfying the initial condition $y(x_0) = y_0$.
Then from (2.2) it follows that

$$y(x) - z(x) = \int_{x_0}^{x} \{f(t, y(t)) - g(t, z(t))\} \, dt$$

We obtain by Assumption 2

$$(2.5) \quad |y(x) - z(x)| = \left| \int_{x_0}^{x} \{f(t, z(t)) - g(t, z(t))\} \, dt \right.$$
$$\left. + \int_{x_0}^{x} \{f(t, y(t)) - f(t, z(t))\} \, dt \right|$$
$$\leq \left| \int_{x_0}^{x} |f(t, z(t)) - g(t, z(t))| \, dt \right|$$
$$+ K \left| \int_{x_0}^{x} |y(t) - z(t)| \, dt \right|$$
$$\leq \varepsilon |x - x_0| + K \left| \int_{x_0}^{x} |y(t) - z(t)| \, dt \right|$$

Therefore, setting

$$\sup_{|x-x_0| \leq h} |y(x) - z(x)| = M'$$

we have

$$|y(x) - z(x)| \leq \varepsilon |x - x_0| + KM' |x - x_0|$$

for $|x - x_0| \leq h$. Substituting this estimate for $|y(t) - z(t)|$ on the
right side of (2.5), we obtain

$$|y(x) - z(x)| \leq M' \frac{K^2 |x - x_0|^2}{2!} + \varepsilon \sum_{m=1}^{2} \frac{K^{m-1} |x - x_0|^m}{m!}$$

for $|x - x_0| \leq h$. Repeating this substitution, we obtain, for each
$n = 1, 2, 3, \cdots,$

$$|y(x) - z(x)| \leq M' \frac{K^n |x - x_0|^n}{n!} + \varepsilon \sum_{m=1}^{n} \frac{K^{m-1} |x - x_0|^m}{m!}$$

for $|x - x_0| \leq h$. As $n \to \infty$, the first term on the right side converges to zero uniformly on the interval $|x - x_0| \leq h$. The second term is less than

$$\varepsilon K^{-1}\{\exp(K|x - x_0|) - 1\}$$

Accordingly, the estimate of the error of the approximate solution $z(x)$ in the interval $|x - x_0| \leq h$ is given by

$$(2.6) \qquad |y(x) - z(x)| \leq (\varepsilon/K)(\exp(K|x - x_0|) - 1)$$

EXAMPLE. Consider the initial value problem

$$dy/dx = \sin(xy)$$

with the initial condition $y(0) = 0.1$. We shall calculate the estimate of the error of the approximate solution $z(x) = 0.1 \exp(\frac{1}{2}x^2)$ which is the solution of the equation

$$dy/dx = xy$$

satisfying the initial condition $z(0) = 0.1$. We take the domain

$$|x| \leq \tfrac{1}{2}, \qquad |y - 0.1| \leq \tfrac{1}{2}$$

as D, so that $a = b = \frac{1}{2}$. Then we have $M = \sup_{(x,y)\in D} |\sin(xy)| < 1$. Moreover, by the mean-value theorem, we have

$$|\sin(xy_1) - \sin(xy_2)| \leq |xy_1 - xy_2|$$

Hence we may set the Lipschitz constant $K = \frac{1}{2}$. Since $a = b = \frac{1}{2}$, and $M < 1$, we have $h = \min(a, b/M) = \frac{1}{2}$. Moreover $|z(x)| \leq \frac{1}{2}$ for $|x| \leq h = \frac{1}{2}$. Hence by the Taylor expansion theorem we obtain

$$|f(x, z(x)) - g(x, z(x))| \leq |\sin(x z(x)) - x z(x)| \leq \frac{|x z(x)|^3}{6} \leq \tfrac{1}{384}$$

for $|x| \leq \frac{1}{2}$. Setting $\varepsilon = \tfrac{1}{384}$ and $K = \frac{1}{2}$ in (2.6), we obtain

$$|y(x) - (0.1)\exp(\tfrac{1}{2}x^2)| \leq \tfrac{2}{384}\left\{\exp\left(\frac{|x|}{2}\right) - 1\right\} \leq \tfrac{0.6}{192} |x|$$

as the estimate of the error of the approximate solution $z(x)$.

3. Integration constants

As was shown in Part 1, a particular initial condition (1.5') determines uniquely the solution $y = \varphi(x)$ of the equation (1.2). For this reason, the solution y is called a *particular solution* of (1.2). In order to indicate the initial condition (1.5'), we denote this solution by

$$(3.1) \qquad y = \varphi(x; x_0, y_0)$$

We now regard the pair (x_0, y_0) as parameters which vary arbitrarily in the region for which Assumptions 1 and 2 hold. Then (3.1), as a function of x containing the parameters x_0, y_0 in the above sense, is called the *general solution* of (1.2). It should be noted that, according to the proof of the existence theorem, the solution (3.1) as a function of x is defined only on the interval $|x - x_0| \leq h$, where

$$h = \min(a, b/M)$$

Both a and b are given in Assumption 1, hence, in general, they depend on both x_0 and y_0. Moreover, according to its definition (1.10), M may also depend on both x_0 and y_0. Therefore the existence domain of the solution (3.1) as a function of x depends on both x_0 and y_0.

For example, consider the differential equation

$$(3.2) \qquad dy/dx = y^2$$

The general solution of (3.2) is given by

$$\varphi(x; x_0, y_0) = \frac{1}{x_0 - x + (1/y_0)}$$

As will be shown later, in the case of *linear* differential equations of the first order, the existence domain under consideration depends only on x_0, but not on y_0. This fact can be extended to the case of linear differential equations of the nth order and because of this fact the theory of linear differential equations has been developed particularly.

The general solution (3.1) seems to depend on two parameters x_0 and y_0; however, essentially, it depends on only one parameter. This will be proved as follows.

Let $f(x, y)$ on the right side of (1.2) satisfy Assumptions 1 and 2. If

$$|x_1 - x_0| \leq \kappa, \qquad |y_1 - y_0| \leq \kappa$$

then $f(x, y)$ is continuous on a domain $D' \subset D$ given by

$$|x - x_1| \leq a - \kappa, \qquad |y - y_1| \leq b - \kappa$$

and moreover, $|f(x, y)| \leq M$ in D'. Moreover, the Lipschitz condition with respect to y holds for $f(x, y)$ on D' with K as the Lipschitz constant. Therefore, according to the existence theorem, the solution $y = \varphi(x; x_1, y_1)$ of (1.2) with $y(x_1) = y_1$ does exist for

$$|x - x_1| \leqq h'_\kappa = \min \left(a - \kappa, \frac{b - \kappa}{M} \right)$$

Accordingly, if the positive number κ is so small that $\kappa < h'_\kappa$, the existence domain of $\varphi(x; x_1, y_1)$ as a function of x contains the point x_0. We write

(3.3) $$\bar{y} = \varphi(x_0; x_1, y_1)$$

Then, by the uniqueness of the solution, we must have

(3.4) $$\varphi(x; x_1, y_1) = \varphi(x; x_0, \bar{y})$$

Therefore, as long as (x_1, y_1) varies in a sufficiently small neighbourhood of the point (x_0, y_0) of D, we may consider x_0 to be fixed and the general solution $\varphi(x; x_1, y_1)$ to be written in a form containing only one parameter C,

$$y = \varphi(x; x_0, C)$$

Since x_0 is fixed, we may write this simply as

$$y = \varphi(x, C)$$

This parameter C is called the *integration constant* of the general solution of the equation (1.2). The general solution of (3.2), for example, may be written as

$$\varphi(x, C) = \frac{1}{C - x}$$

4. *Solution by power series expansion*

We consider the differential equation

(4.1) $$dy/dx = f(x, y)$$

in the case when $f(x, y)$ is a complex-valued function of complex variables x and y. We make the following assumption concerning $f(x, y)$.

ASSUMPTION 1'. The function $f(x, y)$ can be expanded in a convergent power series in $(x - x_0)$ and $(y - y_0)$ in a domain D' of the complex (x, y)-space given by

$$|x - x_0| < a', \qquad |y - y_0| < b'$$

In other words, $f(x, y)$ is a *regular* function in the domain D'.

From this assumption, it follows that $\partial f(x, y)/\partial y$ is also regular in D'. Therefore, for any positive numbers a, b such that $a < a'$ and $b < b'$, both $|f(x, y)|$ and $|\partial f(x, y)/\partial y|$ are continuous on the closed complex domain D given by

$$|x - x_0| \leqq a, \qquad |y - y_0| \leqq b$$

Thus there exist positive numbers M and K such that

(4.2)
$$\sup_{(x, y) \in D} |f(x, y)| = M < \infty$$
$$\sup_{(x, y) \in D} \left| \frac{\partial f(x, y)}{\partial y} \right| = K < \infty$$

Integrating $\partial f(x, y)/\partial y$ along the segment connecting y_1 and y_2, we obtain

$$f(x, y_1) - f(x, y_2) = \int_{y_1}^{y_2} \frac{\partial f(x, y)}{\partial y} dy$$

Hence the Lipschitz condition

(4.3)
$$|f(x, y_2) - f(x, y_1)| \leqq K |y_2 - y_1|$$

holds on D. Therefore, under Assumption 1', we can apply to the equation (4.1) the method of successive approximations on the domain

(4.4)
$$|x - x_0| \leqq h, \qquad h = \min(a, b/M)$$

as follows. We write

$$y_1(x) = y_0 + \int_{x_0}^{x} f(t, y_0) dt$$
$$y_2(x) = y_0 + \int_{x_0}^{x} f(t, y_1(t)) dt$$
$$\cdots\cdots\cdots\cdots\cdots\cdots\cdots\cdots\cdots$$
$$y_n(x) = y_0 + \int_{x_0}^{x} f(t, y_{n-1}(t)) dt$$

where the integration means complex integration along a smooth curve connecting x_0 and x in the domain (4.4). Since $f(x, y_0)$ is regular in the domain $|x - x_0| < h$, the first integral is well-defined, independent of the curves, and hence, so is y_1. . Taking the first integral along the segment connecting x_0 and x, we obtain

$$|y_1(x) - y_0| \leqq hM \leqq b$$

Hence $f(x, y_1(x))$ is well-defined for $|x - x_0| < h$ as a function of x.

Since $y_1(x)$ is given by the integral of the regular function $f(x, y_0)$, $y_1(x)$ is regular in the domain $|x - x_0| < h$. Hence $f(x, y_1(x))$ is also regular. Therefore the second integral is well defined and hence $y_2(x)$ is well defined and regular in the domain $|x - x_0| < h$. Taking the integral along the segment connecting x_0 and x, we obtain further

$$|y_2(x) - y_0| \leqq hM \leqq b$$

In this way, we can define $y_3(x), y_4(x), \cdots$, successively in the domain $|x - x_0| < h$. The functions $y_n(x)$, $n = 1, 2, \cdots$, are all regular in the domain $|x - x_0| < h$ and

$$|y_n(x) - y_0| \leqq b$$

Taking the integrals along the segment connecting x_0 and x and by the same calculation as in Part 1, we can prove that the sequence of regular functions $\{y_n(x)\}$ converges uniformly in the domain $|x - x_0| < h$ and that the limit function $y(x)$ satisfies

$$y(x_0) = y_0 , \qquad \frac{dy(x)}{dx} = f(x, y)$$

in the domain $|x - x_0| < h$. Furthermore, $y(x)$ being the uniform limit of the sequence of regular functions is also regular. The uniqueness of the solution can be proved in the same way as in Part 1.

The method of undetermined coefficients. Since the existence and the uniqueness of the regular solution $y(x)$ are guaranteed, we can calculate this solution by the method of undetermined coefficients as follows. By virtue of its regularity, $y(x)$ can be expanded in a power series

$$y(x) = y_0 + (x - x_0) \left(\frac{dy}{dx} \right)_{x_0} + \frac{(x - x_0)^2}{2!} \left(\frac{d^2y}{dx^2} \right)_{x_0} + \cdots$$

in the domain $|x - x_0| < h$. Substituting this expansion for y on the right side of (4.1), and differentiating both sides, we obtain

$$\frac{dy}{dx} = f(x, y)$$

$$\frac{d^2y}{dx^2} = \frac{\partial f(x, y)}{\partial x} + \frac{\partial f(x, y)}{\partial y} \frac{dy}{dx}$$

$$\frac{d^3y}{dx^3} = \frac{\partial^2 f(x, y)}{\partial x^2} + 2 \frac{\partial^2 f(x, y)}{\partial x \partial y} \frac{dy}{dx} +$$

$$+ \frac{\partial^2 f(x, y)}{\partial y^2} \left(\frac{dy}{dx} \right)^2 + \frac{\partial f(x, y)}{\partial y} \frac{d^2 y}{dx^2}$$

. . . .

Setting in these relations $x = x_0$ and $y = y_0$, we can determine successively the expansion coefficients $(dy/dx)_{x_0}, (d^2y/dx^2)_{x_0}, \cdots$.

Picard's remark on the uniqueness of the solution. Let C be a curve in the complex x-plane such that the point x_0 is an end point of C. Let $y(x)$ satisfy the equation (4.1) and be regular at any point on C except the point x_0. Let $y(x)$ tend to y_0 as x tends to x_0 along C. If Assumption $1'$ is satisfied in the complex domain D', then $y(x)$ is also regular at the point x_0 and this solution coincides with the unique regular solution of (4.1) satisfying the initial condition $y(x_0) = y_0$. Owing to this fact, the study of the analytic continuation of regular solutions can be developed smoothly and for this reason, Picard's remark plays a basic role in the study of the solutions from a function theoretical standpoint of view.

Painlevé's proof of Picard's remark. From the assumption it follows that, for any $\varepsilon > 0$, there exists a point x_1 on C such that

$$|x_1 - x_0| < \varepsilon, \qquad |y(x_1) - y_0| < \varepsilon$$

Let D'' be a complex domain given by

$$|x - x_1| \leq a - \varepsilon, \qquad |y - y(x_1)| \leq b - \varepsilon$$

Then, for sufficiently small ε, D'' is contained in D. Hence by (4.2),

$$\sup_{(x,y) \in D''} |f(x, y)| \leq M$$

Accordingly, as was shown in Part 3, there exists a unique regular solution $\varphi(x; x_1, y(x_1))$ of (4.1) satisfying the initial condition $y = y(x_1)$ at $x = x_1$ in the domain

(4.5) $$|x - x_1| < h(x_1) = \min \left(a - \varepsilon, \frac{b - \varepsilon}{M} \right)$$

Since

$$\lim_{x_1 \to x_0} h(x_1) = \lim_{\varepsilon \to 0} \left(a - \varepsilon, \frac{b - \varepsilon}{M} \right)$$

$$= \min \left(a, \frac{b}{M} \right) = h$$

the domain (4.5), in which $\varphi(x; x_1, y(x_1))$ is regular, contains the point x_0, if ε is sufficiently small such that

$$\min\left(a - \varepsilon, \frac{b - \varepsilon}{M}\right) > \varepsilon$$

On the other hand, both $\varphi(x; x_1, y(x_1))$ and $y(x)$ are regular at the point x_1 and further $\varphi(x_1; x_1, y(x_1)) = y(x_1)$. Therefore, by the uniqueness of the regular solution, we must have

$$\varphi(x; x_1, y(x_1)) \equiv y(x)$$

in the domain (4.5). Thus $y(x)$ is also regular at the point x_0.

REMARK. It should be noted that the above proof is based upon the same idea by which we derived (3.4) in Part 3. Accordingly, we can prove, similarly, the following. Let $f(x, y)$ satisfy Assumption 1 and 2 in Part 1 in the domain D of the (x, y)-plane given by

$$|x - x_0| \leqq a, \qquad |y - y_0| \leqq b$$

Let I be an interval with x_0 as an end point. If $y(x)$ satisfies the equation $dy/dx = f(x, y)$ at any point in I except the point x_0, and if $y(x)$ tends to y_0 as x tends to x_0 along I, then, in some neighbourhood of the point $x_0, y(x)$ coincides with the unique continuous solution $z(x)$ of this equation satisfying the initial condition $z(x_0) = y_0$.

5. *Differential equations containing parameters.*
Perturbation theory.

We consider the differential equation

(5.1) $dy/dx = f(x, y, \lambda)$

containing a parameter λ. The *perturbation theory* deals with the problem of finding an approximate solution of (5.1), for sufficiently small $|\lambda - \lambda_0|$, under the initial condition

(5.2) $y = y_0 \qquad \text{at} \quad x = x_0$

when we know, for $\lambda = \lambda_0$, the solution of (5.1) satisfying the initial condition (5.2). Such parameter λ is called the *perturbation parameter*.

To formulate the problem precisely, we assume that the variables y, λ of $f(x, y, \lambda)$ are complex—x may be either real or complex—and $f(x, y, \lambda)$ is continuous on a domain D_1 given by

$$|x - x_0| \leqq a, \qquad |y - y_0| \leqq b, \qquad |\lambda - \lambda| \leqq c$$

and furthermore

$$\sup |f(x, y, \lambda)| = M < \infty$$

$$\sup \left| \frac{\partial f(x, y, \lambda)}{\partial y} \right| = K < \infty$$

on D_1. Accordingly, the Lipschitz condition

$$|f(x, y, \lambda) - f(x, y_1, \lambda)| \leqq K |y - y_1|$$

holds, independent of λ, on D_1. If, moreover, $f(x, y, \lambda)$ is regular in D_1 as a function of y and λ, then the solution $y(x, \lambda)$ of (5.1) satisfying the initial condition (5.2), which is independent of λ, is regular at the point λ_0 as a function of λ.

Proof. We write $h = \min(a, b/M)$. Similarly as in Part 1, we can define the successive approximate functions $y_n(x, \lambda)$ for $|x - x_0| \leqq h$, $|\lambda - \lambda_0| \leqq c$ as follows,

$$y_1(x, \lambda) = y_0 + \int_{x_0}^{x} f(t, y_0, \lambda) \, dt$$

$$y_2(x, \lambda) = y_0 + \int_{x_0}^{x} f(t, y_1(t, \lambda), \lambda) \, dt$$

$$\cdots\cdots\cdots\cdots\cdots\cdots\cdots$$

$$y_n(x, \lambda) = y_0 + \int_{x_0}^{x} f(t, y_{n-1}(t, \lambda), \lambda) \, dt$$

Furthermore, we can prove that $\{y_n(x, \lambda)\}$ converges uniformly and the limit $y(x)$ satisfies (5.1) and (5.2) on the domain $|x - x_0| \leqq h$, $|\lambda - \lambda_0| \leqq c$. On the other hand, by our assumption, $y_1(x, \lambda)$ is regular in the domain $|\lambda - \lambda_0| < c$ as a function of λ; for $y_1(x, \lambda)$ is differentiable with respect to the complex variable λ, namely,

$$\frac{\partial y_1(x, \lambda)}{\partial \lambda} = \int_{x_0}^{x} \frac{\partial f(t, y_0, \lambda)}{\partial \lambda} \, dt$$

In the same manner, we can prove that $y_2(x, \lambda), y_3(x, \lambda), \cdots$, are all regular in the domain $|\lambda - \lambda_0| < c$ as a function of λ. Therefore, being the uniform limit of regular functions, $y(x, \lambda)$ is regular in the domain $|\lambda - \lambda_0| < c$ as a function of λ.

Variation equation. Since $y(x, \lambda)$ is regular as a function of λ, $y(x, \lambda)$ can be expanded in a power series in $(\lambda - \lambda_0)$ as follows

$$(5.3) \qquad y(x, \lambda) = y(x, \lambda_0) + \sum_{n=1}^{\infty} (\lambda - \lambda_0)^n z_n(x)$$

For sufficiently small $|\lambda - \lambda_0|$, we have approximately

$$y(x, \lambda) = y(x, \lambda_0) + (\lambda - \lambda_0)z_1(x)$$

where $z_1(x)$ is given by

$$\left(\frac{\partial y(x, \lambda)}{\partial \lambda} \right)_{\lambda=\lambda_0}$$

Substituting $y(x, \lambda)$ for y in (5.1), and differentiating both sides with respect to λ, we obtain, by setting $\lambda = \lambda_0$,

$$(5.4) \qquad \frac{dz_1}{dx} = \left(\frac{\partial f(x, y(x, \lambda), \lambda)}{\partial \lambda} \right)_{\lambda=\lambda_0}$$

$$= z_1 \left(\frac{\partial f(x, Y, \lambda)}{\partial Y} \right)_{\substack{\lambda=\lambda_0 \\ Y=y(x,\lambda_0)}} + \left(\frac{\partial f(x, Y, \lambda)}{\partial \lambda} \right)_{\substack{\lambda=\lambda_0 \\ Y=y(x,\lambda_0)}}$$

Thus, $z_1(x)$ satisfies a linear differential equation of the first order with the coefficients depending on x, $y(x, \lambda_0)$ and λ_0. The equation (5.4) gives us the principal term $(\lambda - \lambda_0)z_1(x)$ of the variation $y(x, \lambda) - y(x, \lambda_0)$ for sufficiently small $|\lambda - \lambda_0|$. This is the reason why Poincaré called this equation "l'équation aux variation." By (5.2) and (5.3), we obtain

$$\sum_{n=1}^{\infty} (\lambda - \lambda_0)^n z_n(x_0) \equiv 0$$

Hence the initial condition (5.2) yields the initial condition $z_1(x_0) = 0$ for the variation equation (5.4). Solving the equation (5.4) under this initial condition $z_1(x_0) = 0$, we obtain

$$(5.5) \qquad z_1(x) = \left\{ \int_{x_0}^{x} q(t) \exp \left(\int_{x_0}^{t} -p(\tau) \, d\tau \right) dt \right\} \exp \left(\int_{x_0}^{x} p(t) \, dt \right)$$

where

$$p(x) = \left(\frac{\partial f(x, Y, \lambda)}{\partial Y} \right)_{\substack{\lambda=\lambda_0 \\ Y=y(x,\lambda_0)}} , \qquad q(x) = \left(\frac{\partial f(x, Y, \lambda)}{\partial \lambda} \right)_{\substack{\lambda=\lambda_0 \\ Y=y(x,\lambda_0)}}$$

REMARK. The method of obtaining (5.5) is as follows. In

$$(5.6) \qquad dz/dx = p(x)z + q(x)$$

put

$$(5.7) \qquad z = wy , \qquad y(x) = \exp \left(\int_{x_0}^{x} p(t) \, dt \right)$$

Then (5.6) becomes

$$y \frac{dw}{dx} = q$$

The solution of this equation is given by

$$w(x) = \int_{x_0}^{x} \frac{q(t)}{y(t)} \, dt + C$$

Hence the general solution of (5.6) is given by

$$(5.8) \quad z(x) = \left\{ \int_{x_0}^{x} q(t) \exp\left(-\int_{x_0}^{t} p(\tau) \, d\tau \right) dt + C \right\} \exp\left(\int_{x_0}^{x} p(t) \, dt \right)$$

where C is the constant of integration.

Regularity of the general solution with respect to the initial condition. We consider again the equation

$$\frac{dy}{dx} = f(x, y)$$

If the function $f(x, y)$ is regular in a domain of the complex (x, y)-space given by

$$|x - x_0| < a', \qquad |y - y_0| < b'$$

then the general solution $y = \varphi(x, x_0, y_0)$ is regular as a function of x_0 and y_0. To prove this, set

$$x = x_0 + \xi, \qquad y = y_0 + \eta$$

Then the equation considered becomes

$$\frac{d\eta}{d\xi} = f(x_0 + \xi, y_0 + \eta) = F(\xi, \eta, x_0, y_0)$$

Considering x_0, y_0 as parameters, we can also apply the preceding discussion to the solution of this equation satisfying the initial condition $\eta = 0$ at $\xi = 0$.

6. Existence and uniqueness of the solution of a system of differential equations

An nth order ordinary differential equation is generally written in the form

$$F\left(x, y, \frac{dy}{dx}, \frac{d^2 y}{dx^2}, \cdots, \frac{d^n y}{dx^n} \right) = 0$$

We shall restrict ourselves to those cases for which the equation can be solved for $d^n y/dx^n$ and written in the form

$$(6.1) \qquad \frac{d^n y}{dx^n} = f\left(x, y, \frac{dy}{dx}, \cdots, \frac{d^{n-1} y}{dx^{n-1}} \right)$$

In order to solve the equation (6.1) under the initial conditions

$$(6.2) \qquad y(x_0) = y_0, \ \frac{dy}{dx}(x_0) = y_0', \ \cdots, \ \frac{d^{n-1}y}{dx^{n-1}}(x_0) = y_0^{(n-1)}$$

it is sufficient to solve the system of first-order differential equations

$$\frac{dy}{dx} = y_1$$

$$\frac{dy_1}{dx} = y_2$$

$$(6.3) \qquad \cdots\cdots\cdots$$

$$\frac{dy_{n-2}}{dx} = y_{n-1}$$

$$\frac{dy_{n-1}}{dx} = f(x, y, y_1, \cdots, y_{n-1})$$

with the initial conditions

$$(6.4) \qquad y(x_0) = y_0, \quad y_1(x_0) = y_0', \quad y_2(x_0) = y_0'', \cdots, \quad y_{n-1}(x_0) = y_0^{(n-1)}$$

We may consider, more generally, the system of ordinary differential equations

$$\frac{dz_1}{dx} = f_1(x, z_1, z_2, \cdots, z_n)$$

$$\frac{dz_2}{dx} = f_2(x, z_1, z_2, \cdots, z_n)$$

$$(6.5) \qquad \cdots\cdots\cdots\cdots\cdots$$

$$\frac{dz_n}{dx} = f_n(x, z_1, z_2, \cdots, z_n)$$

with the initial conditions

$$(6.6) \qquad z_m(x_0) = y_0^{(m-1)}, \qquad m = 1, 2, \cdots, n$$

where $y_0^{(0)} = y_0$. For this problem we shall prove the following
  THEOREM 6.1. Let

$$f_1(x, z_1, \cdots, z_n), f_2(x, z_1, \cdots, z_n), \cdots, f_n(x, z_1, \cdots, z_n)$$

be real-valued and continuous on a domain D of the real $(x, z_1, \cdots, z_n)$-space given by

$$(6.7) \qquad |x - x_0| \leqq a, \qquad |z_m - y_0^{(m-1)}| \leqq b, \qquad m = 1, 2, \cdots, n$$

Assume that the Lipschitz condition with respect to $z_1, z_2, \cdots, z_n$ is satisfied in D, that is, there exists a positive constant K such that, for every pair of points $(x, \zeta_1, \cdots, \zeta_n), (x, \eta_1, \cdots, \eta_n)$ in D,

$$|f_i(x, \zeta_1, \cdots, \zeta_n) - f_i(x, \eta_1, \cdots, \eta_n)| \leq K \sum_{m=1}^{n} |\zeta_m - \eta_m|$$

for every $i = 1, 2, \cdots, n$. Further let

(6.8)
$$h = \min(a, b/M)$$
$$M = \sup_{\substack{(x, z_1, \cdots, z_n) \in D \\ i = 1, 2, \cdots, n}} |f_i(x, z_1, \cdots, z_n)|$$

Then there exists one and only one set of solutions $z_1(x), z_2(x), \cdots, z_n(x)$ of (6.5) on the interval

(6.9)
$$|x - x_0| \leq h$$

satisfying the initial conditions (6.6).

This theorem implies immediately the following

THEOREM 6.2. Assume that $f(x, z_1, z_2, \cdots, z_n)$ is real-valued and continuous on the domain D and satisfies the Lipschitz condition on D, that is, for every pair of points $(x, \zeta_1, \cdots, \zeta_n), (x, \eta_1, \cdots, \eta_n)$ of D,

$$|f(x, \zeta_1, \cdots, \zeta_n) - f(x, \eta_1, \cdots, \eta_n)| \leq K \sum_{m=1}^{n} |\zeta_m - \eta_m|$$

Then there exists one and only one solution $y(x)$ of the equation (6.1) satisfying the initial conditions (6.2) on the interval

$$|x - x_0| \leq h$$

where $h = \min(a, b/M)$ and $M = \sup_{(x, z_1, \cdots, z_n) \in D} |f(x, z_1, z_2, \cdots, z_n)|$.

Proof of Theorem 6.1. The proof is entirely the same as in the case of the first order equation in Part 1. The initial value problem for (6.5) with (6.6) can be reduced to the system of integral equations

$$z_m(x) = y_0^{(m-1)} + \int_{x_0}^{x} f_m(t, z_1(t), \cdots, z_n(t)) dt \qquad (m = 1, 2, \cdots, n)$$

and solved by the method of successive approximations. In this case, the successive approximate functions are defined by

$$z_{m,1}(x) = y_0^{(m-1)} + \int_{x_0}^{x} f_m(t, y_0, y_0', \cdots, y_0^{(n-1)}) dt$$

$$z_{m,2}(x) = y_0^{(m-1)} + \int_{x_0}^{x} f_m(t, z_{1,1}(t), z_{2,1}(t), \cdots, z_{n,1}(t)) dt$$

$$\cdots\cdots\cdots\cdots\cdots\cdots\cdots$$

$$z_{m,k}(x) = y_0^{(m-1)} + \int_{x_0}^{x} f_m(t, z_{1,k-1}(t), z_{2,k-1}(t), \cdots, z_{n,k-1}(t)) dt$$

Then, by virtue of the Lipschitz condition, we obtain

$$\sum_{m=1}^{n} |z_{m,k}(x) - z_{m,k-1}(x)| \leqq K \left| \int_{x_0}^{x} \sum_{m=1}^{n} |z_{m,k-1}(t) - z_{m,k-2}(t)| dt \right|$$

From this, we obtain, in the same manner as in Part 1, that, for $k > s$,

$$(6.10) \qquad \sum_{m=1}^{n} |z_{m,k}(x) - z_{m,s}(x)| \leqq nb \sum_{t=s}^{k-1} \frac{(K|x - x_0|)^t}{t!}$$

on the interval (6.9), provided that $z_{m,0}(x) = y_0^{(m-1)}$. This suffices to prove the theorem.

Similarly, it is easy to prove the following theorems.

THEOREM 6.3. If $f_1(x, z_1, \cdots, z_n), \cdots, f_n(x, z_1, \cdots, z_n)$ are complex-valued functions of complex variables $x, z_1, \cdots, z_n$, regular in the interior of the complex domain given by (6.7), then there exists one and only one set of solutions $z_1(x), z_2(x), \cdots, z_n(x)$ of (6.5) in the domain $|x - x_0| < h$ satisfying the initial conditions (6.6). Further $z_1(x), z_2(x), \cdots, z_n(x)$ are all regular for $|x - x_0| < h$.

THEOREM 6.4. If $f(x, z_1, \cdots, z_n)$ is a complex-valued function of complex variables $x, z_1, \cdots, z_n$ regular in the interior of the complex domain given by (6.7), then there exists one and only one solution $y(x)$ of (6.1) satisfying the initial conditions (6.2) in the domain $|x - x_0| < h$, where $h = \min(a, b/M)$ and $M = \sup_{(x, z_1, \ldots, z_n) \in D} \cdot |f(x, z_1, \cdots, z_n)|$. Further $y(x)$ is regular in the domain $|x - x_0| < h$.

We can also calculate the power series expansions of these solutions by the method of undetermined coefficients as in Part 4.

It is not difficult to prove the following theorems, which are the extensions of Picard's remark mentioned in Part 4.

THEOREM 6.5. Let the hypothesis of Theorem 6.3 be satisfied. Let C be a curve in the complex x-plane with x_0 as one of its end points. Let $z_1(x), z_2(x), \cdots, z_n(x)$ satisfy (6.5) and be regular at any point on

C except the point x_0. If $z_m(x)$, $m = 1, 2, \cdots, n$, tends to $y_0^{(m-1)}$ as x tends to x_0 along C, then $z_m(x)$, $m = 1, 2, \cdots, n$, is also regular at the point x_0.

THEOREM 6.6. Let the hypothesis of Theorem 6.4 be satisfied. Let C be a curve as in Theorem 6.5. Let $y(x)$ satisfy (6.1) and be regular at any point on C except the point x_0. If $y^{(m)}(x)$, $m = 0, 1, 2, \cdots, n - 1$, tends to $y_0^{(m)}$ where $y^{(0)}(x) = y(x)$, $y_0^{(0)} = y_0$, then $y(x)$ is also regular at the point x_0.

THEOREM 6.7. Let the hypothesis of Theorem 6.2 be satisfied. Let C be an interval with x_0 as one of its end points. If $y(x)$ satisfies (6.1) at any point on C except the point x_0, and if

$$y(x) \to y_0, y'(x) \to y_0', \cdots, y^{(n-1)}(x) \to y_0^{(n-1)}$$

as $x \to x_0$ along C, then, in some neighbourhood of x_0, $y(x)$ coincides with the unique continuous solution of (6.1) satisfying the initial conditions (6.2).

Finally, we remark that the perturbation theory of nth order differential equations or systems of differential equations containing parameters can be discussed in the same manner as in Part 5.

§2. Linear differential equations of the nth order

7. Singular points for linear differential equations

A system of differential equations (6.5) is said to be *linear* when $f_m(x, z_1, z_2, \cdots, z_n)$, $m = 1, 2, \cdots, n$, is of the form

$$(7.1) \qquad f_m(x, z_1, z_2, \cdots, z_n) = \sum_{i=1}^{n} a_{m,i}(x) z_i + b_m(x)$$

We consider a linear system of first order equations

$$(7.2) \qquad dz_m/dx = \sum_{i=1}^{n} a_{m,i}(x) z_i + b_m(x) \qquad (m = 1, 2, \cdots, n)$$

where the coefficients $a_{m,i}(x)$, $b_m(x)$ are real-valued (or complex-valued) functions on a real (or complex) domain D given by

$$(7.3) \qquad |x - x_0| \leqq a$$

If all the coefficients $a_{m,i}(x)$, $b_m(x)$ are continuous on (or regular on) D, then, for any initial values $y_0, y_0', \cdots, y_0^{(n-1)}$, there exists one and only one set of solutions $z_1(x), z_2(x), \cdots, z_n(x)$ of (7.2) on D, satisfying the initial conditions

(7.4) $z_m(x_0) = y_0^{(m-1)}$ $(m = 1, 2, \cdots, n;\ y_0^{(0)} = y_0)$

Further, $z_1(x),\ z_2(x),\ \cdots,\ z_n(x)$ are continuous on (or regular in) D.

Proof. It is easily seen that the functions on the right side of (7.2) satisfy the Lipschitz condition with respect to $z_1, z_2, \cdots, z_n$, in a domain

(7.5) $|x - x_0| \leqq a,\ \ |z_m| < \infty$ $(m = 1, 2, \cdots, n)$

Hence we may prove the existence and uniqueness of solutions by the method of successive approximations as in Part 6. In Part 6, we restrict the domain of successive approximation functions $z_{m,i}(x)$ to $|x - x_0| \leqq h < a$, for the reason that it can not be guaranteed that the values of $z_{m,i}(x)$ always fall into the domain (6.7) of $f_1, f_2, f_3, \cdots, f_n$. In the present case, however, such a restriction is no more needed, as is easily seen from (7.5). Hence, we can apply the method of successive approximations on the whole domain D, and obtain the result mentioned above, q.e.d.

REMARK 1. If $f_m(x, z_1, \cdots, z_n)$ is not linear, but a polynomial in $z_1, z_2, \cdots, z_n$, then the procedure mentioned above is not always available; for the Lipschitz condition would not hold in the whole domain (7.5). For example, consider the equation

$$dz/dx = z^2,\qquad z(0) = 1$$

The solution is given by $z(x) = 1/(1 - x)$. Obviously $z(x)$ has a singularity at the point $x = 1$, while the coefficients of the equation have no singularity at this point.

REMARK 2. Some of the solutions of a linear differential equation are continuous at the singular point of the coefficients. For example, consider the function $y = 2x$. Obviously, y satisfies the equation

$$\frac{dy}{dx} = \frac{1}{x}y$$

and is continuous (regular) at the point $x = 0$, while the coefficient $1/x$ has a discontinuity (a simple pole) at this point.

REMARK 3. Let $y(x)$ be the solution of the initial value problem for the nth order linear differential equation

(7.6) $$\frac{d^n y}{dx^n} = \sum_{i=0}^{n-1} a_i(x)\frac{d^i y}{dx} + b(x)$$

with initial conditions

(7.7) $y(x_0) = y_0,\ y'(x_0) = y_0',\ \cdots,\ y^{(n-1)}(x_0) = y_0^{(n-1)}$

If the coefficients $a_i(x)$ and $b(x)$ are continuous on (or regular on) the domain (7.3), then, $y(x)$, together with its derivatives $y'(x), y''(x)$, $\cdots, y^{(n-1)}(x)$, is continuous on (or regular in) the domain (7.3). This is easily derived from the above result, by reducing (7.6) to a linear system as (6.3).

REMARK 4. If at least one of the coefficients of (7.2) has a discontinuity (or singularity) at a point x, then the point x is called a *singular point* for the system (7.2).

According to the result·mentioned above, the solutions of a linear differential equation or a linear system have no discontinuity (or singularity) at any point except at the singular points for the equation or the system.

8. *Fundamental system of solutions*

A linear ordinary differential equation of the nth order

$$(8.1) \qquad p_0(x)y^{(n)} + p_1(x)y^{(n-1)} + \cdots + p_{n-1}(x)y' + p_n(x)y = q(x)$$

is called *homogeneous* when $q(x) \equiv 0$, and *inhomogeneous* when $q(x) \not\equiv 0$. We assume that the coefficients $p_0(x), p_1(x), \cdots, p_n(x)$ and $q(x)$ are all continuous on a domain D and further that $p_0(x) \neq 0$ on D. On the domain D, we may divide the both sides of (8.1) by $p_0(x)$. Therefore, we may consider the equation (8.1) to be written in the form

$$(8.2) \qquad y^{(n)} + p_1(x)y^{(n-1)} + \cdots + p_{n-1}(x)y' + p_n(x)y = q(x)$$

In the following, we shall be concerned with the equation (8.2) under the assumption that the coefficients are all continuous in D. The result in Part 7 implies at once the following

THEOREM 8.1. For any point x_0 in D and for any arbitrary set of n numbers $\eta, \eta', \cdots, \eta^{(n-1)}$, there exists one and only one solution $y(x)$ of the equation (8.2) in D satisfying the initial conditions

$$(8.3) \qquad y(x_0) = \eta, \, y'(x_0) = \eta', \cdots, y^{(n-1)}(x_0) = \eta^{(n-1)}$$

Further, the solution $y(x)$, together with its derivatives $y'(x), y''(x)$, $\cdots, y^{(n)}(x)$, is continuous in D.

COROLLARY. If a solution $y(x)$ of the homogeneous equation

$$(8.4) \qquad y^{(n)} + p_1(x)y^{(n-1)} + \cdots + p_{n-1}(x)y' + p_n(x)y = 0$$

satisfies

$$(8.5) \qquad y(x_0) = y'(x_0) = \cdots = y^{(n-1)}(x_0) = 0$$

at a point x_0 in D, then $y(x)$ is identically zero.

Proof. The function which is identically zero certainly satisfies (8.4) and (8.5). Therefore, according to the uniqueness of the solution, $y(x)$ must be identically zero.

Theorem 8.2. Let $y_1, y_2, \cdots, y_m$ be solutions of the homogeneous equation (8.4). Then any linear combination

$$(8.6) \qquad y = \sum_{i=1}^{m} C_i y_i$$

of these solutions, with arbitrary coefficients $C_1, C_2, \cdots, C_m$, is also a solution of (8.4). This fact is called the *principle of superposition*.

Proof. The proof follows immediately from the fact that

$$y^{(k)} = \sum_{i=1}^{m} C_i y_i^{(k)}$$

for $k = 1, 2, \cdots, n$, q.e.d.

Let $y_1, y_2, \cdots, y_{n+1}$ be an arbitrary set of $n + 1$ solutions of (8.4). Then there exist $n + 1$ numbers $C_1, C_2, \cdots, C_{n+1}$, not all zero, such that

$$\sum_{i=1}^{n+1} C_i y_i(x) \equiv 0$$

To prove this, put $y_i^{(j)}(x_0) = \eta_i^{(j)}$ $(\eta_i^{(0)} = \eta_i)$ and consider the system of n linear equations, in $n + 1$ unknowns $C_1, C_2, \cdots, C_{n+1}$,

$$(8.7) \qquad \sum_{i=1}^{n+1} C_i \eta_i^{(j)} = 0, \qquad j = 0, 1, 2, \cdots, n - 1$$

Since the number n of the equations is less than that of unknowns, there exists a set of $n + 1$ numbers $C_1, C_2, \cdots, C_{n+1}$, not all zero, which satisfies (8.7). Put

$$y(x) = \sum_{i=1}^{n+1} C_i y_i(x)$$

Then $y(x)$ satisfies both (8.4) and (8.5). Therefore, from the corollary of Theorem 8.1, it follows that

$$y(x) = \sum_{i=1}^{n+1} C_i y_i(x) \equiv 0$$

If there exist m numbers $C_1, C_2, \cdots, C_m$, not all zero, such that

$$(8.8) \qquad \sum_{i=1}^{m} C_i y_i(x) \equiv 0$$

in D, then these m functions $y_1(x), y_2(x), \cdots, y_m(x)$, defined in D, are said to be *linearly dependent*. The m functions, which are not linearly dependent, are said to be *linearly independent*.

Theorem 8.3. If $m > n$, m solutions of (8.4) are linearly dependent.

Proof. Let $y_1, y_2, \cdots, y_m$ be a set of m solutions of (8.4). As is shown above, for $n + 1$ solutions $y_1, y_2, \cdots, y_{n+1}$, there exist $n + 1$ numbers $C_1, C_2, \cdots, C_{n+1}$, not all zero, such that $\sum_{i=1}^{n+1} C_i y_i(x) \equiv 0$. Therefore, setting $C_{n+k} = 0$, for $k = 2, 3, \cdots, m - n$, we have

$$\sum_{i=1}^{m} C_i y_i(x) = \sum_{i=1}^{n+1} C_i y_i(x) \equiv 0$$

This means that $y_1, y_2, \cdots, y_m$ are linearly dependent.

THEOREM 8.4. There exist n linearly independent solutions of (8.4).

Proof. Let $y_1(x), y_2(x), \cdots, y_n(x)$ be the solutions of (8.4) satisfying the initial conditions,

(8.9)
$$y_1(x_0) = 1, y_1'(x_0) = 0, \cdots, y_1^{(n-1)}(x_0) = 0$$
$$y_2(x_0) = 0, y_2'(x_0) = 1, \cdots, y_2^{(n-1)}(x_0) = 0$$
$$\cdots\cdots\cdots\cdots\cdots\cdots$$
$$y_n(x_0) = 0, y_n'(x_0) = 0, \cdots, y_n^{(n-1)}(x_0) = 1$$

respectively. Then these solutions are linearly independent. To prove this, let $C_1, C_2, \cdots, C_n$ be an arbitrary set of n numbers, not all zero, say $C_{k+1} \neq 0$, and let

$$y(x) = \sum_{i=1}^{n} C_i y_i(x)$$

Then $y(x)$ is also a solution of (8.4) and satisfies

$$y^{(k)}(x_0) = \sum_{i=1}^{n} C_i y_i^{(k)}(x_0) = C_{k+1} y_{k+1}^{(k)}(x_0) = C_{k+1} \neq 0$$

Hence $y(x) = \sum_{i=1}^{n} C_i y_i(x)$ is not identically zero, that is, the solutions $y_1(x), y_2(x), \cdots, y_n(x)$ are linearly independent.

Definition. A set of n solutions of (8.4) which are linearly independent is called a *fundamental system of the solutions* of (8.4).

For example, the solutions $y_1(x), y_2(x), \cdots, y_n(x)$, defined above, form a fundamental system. Clearly, there are infinity of fundamental systems.

THEOREM 8.5. Let $y_1(x), y_2(x), \cdots, y_n(x)$ be a fundamental system of the solutions of (8.4). Then every solution $y(x)$ can be written uniquely as the linear combination

$$(8.10) \qquad\qquad y(x) = \sum_{i=1}^{n} C_i y_i(x)$$

of $y_1(x), y_2(x), \cdots, y_n(x)$.

Proof. Let $y(x)$ be a solution of (8.4). Then, by Theorem 8.3, $y(x), y_1(x), y_2(x), \cdots, y_n(x)$ are linearly dependent, that is, there exist $n + 1$ numbers $r_1, r_2, \cdots, r_{n+1}$, not all zero, such that

$$r_1 y(x) = \sum_{i=1}^{n} r_{i+1} y_i(x)$$

Since $y_1(x), y_2(x), \cdots, y_n(x)$ are linearly independent, obviously $r_1 \neq 0$. Setting $C_i = r_1^{-1} r_{i+1}$, we obtain (8.10).

To prove the uniqueness of the coefficients $C_1, C_2, \cdots, C_n$, let $y(x)$ be written in the form

$$y(x) = \sum_{i=1}^{n} B_i y_i(x)$$

Then we obtain

$$\sum_{i=1}^{n} (C_i - B_i) y_i(x) \equiv 0$$

Since $y_1(x), y_2(x), \cdots, y_n(x)$ are linearly independent, we must have $C_i = B_i$ for every $i = 1, 2, \cdots, n$.

Relations between the solutions and the coefficients. Let $y_1(x), y_2(x), \cdots, y_n(x)$ be a fundamental system of the solutions of (8.4). If every solution $y_i(x)$ satisfies another equation

$$(8.11) \qquad y^{(n)} + r_1(x) y^{(n-1)} + \cdots + r_{n-1}(x) y' + r_n(x) y = 0$$

with continuous coefficients $r_i(x)$, $i = 1, 2, \cdots, n$, in the domain D, then we have

$$(8.12) \qquad\qquad r_i(x) \equiv p_i(x) , \qquad (i = 1, 2, \cdots, n)$$

Similarly as in the case of algebraic equations, this fact may be stated as follows.

THEOREM 8.6. The coefficients of a linear differential equation of the nth order are determined uniquely by an arbitrary chosen fundamental system of the solutions, provided that the coefficient of $y^{(n)}$ is identically one.

Proof. It is sufficient to prove (8.12). Suppose that (8.12) is not true, namely, there exists an $i \geq 1$ such that

$$r_1(x) \equiv p_1(x) , \qquad r_2(x) \equiv p_2(x), \cdots$$
$$r_{i-1}(x) \equiv p_{i-1}(x) , \qquad r_i(x) \not\equiv p_i(x)$$

Then $y_1(x), y_2(x), \cdots, y_n(x)$ satisfies the equation

$$(8.13) \qquad \{r_i(x) - p_i(x)\}y^{(n-i)} + \cdots + \{r_n(x) - p_n(x)\}y = 0$$

which is the difference of (8.4) and (8.11); for $y_1(x), y_2(x), \cdots, y_n(x)$ satisfies both (8.4) and (8.11). The equation (8.13) is of order $< n$. Hence, by Theorem 8.3, n solutions $y_1(x), y_2(x), \cdots, y_n(x)$ must be linearly dependent in the open domain $r_i(x) \neq p_i(x)$. Thus, by the Corollary of Theorem 8.1, this linear dependence holds in the whole domain D.

9. Wronskian. Liouville's formula

We shall enter into the details of the relations between the solutions and the coefficients mentioned in Part 8. We denote by $W(y, y_1, y_2, \cdots, y_n)$ the determinant

$$\begin{vmatrix} y & y_1 & y_2 & \cdots & y_n \\ y' & y_1' & y_2' & \cdots & y_n' \\ y'' & y_1'' & y_2'' & \cdots & y_n'' \\ \cdots\cdots\cdots\cdots\cdots\cdots\cdots \\ y^{(n)} & y_1^{(n)} & y_2^{(n)} & \cdots & y_n^{(n)} \end{vmatrix}$$

which is called the *Wronskian* of the $n + 1$ functions $y, y_1, y_2, \cdots, y_n$. We consider the linear differential equation

$$(9.1) \qquad W(y, y_1(x), y_2(x), \cdots, y_n(x)) = 0$$

where y is unknown and $y_1(x), y_2(x), \cdots, y_n(x)$ is a fundamental system of the solutions of (8.4). Since

$$W(y_i(x), y_1(x), y_2(x), \cdots, y_n(x)) = 0 \qquad (i = 1, 2, \cdots, n)$$

every $y_i(x)$ satisfies the equation (9.1). Furthermore, as will be shown shortly, the coefficient

$$(9.2) \qquad (-1)^n W(y_1(x), y_2(x), \cdots, y_n(x))$$

of $y^{(n)}$ in (9.1) does not vanish at any point in the domain D. Therefore, by Theorem 8.6, we obtain the following identity

$$(9.3) \qquad y^{(n)} + p_1(x)y^{(n-1)} + \cdots + p_{n-1}(x)y' + p_n(x)y$$
$$\equiv \frac{(-1)^n W(y, y_1(x), y_2(x), \cdots, y_n(x))}{W(y_1(x), y_2(x), \cdots, y_n(x))}$$

This gives the relations between the solutions and the coefficients.

Now we shall prove that (9.2) does not vanish at any point in D. Suppose that there exists a point x_0 in D for which

$$(9.4) \qquad W(y_1(x_0), y_2(x_0), \cdots, y_n(x_0)) = 0$$

Then the system of linear equations with the coefficients $y_i^{(j)}(x_0)$

$$C_1 y_1(x_0) + C_2 y_2(x_0) + \cdots + C_n y_n(x_0) = 0$$
$$C_1 y_1'(x_0) + C_2 y_2'(x_0) + \cdots + C_n y_n'(x_0) = 0$$
$$\cdots\cdots\cdots\cdots\cdots\cdots\cdots\cdots\cdots\cdots\cdots$$
$$C_1 y_1^{(n-1)}(x_0) + C_2 y_2^{(n-1)}(x_0) + \cdots + C_n y_n^{(n-1)}(x_0) = 0$$

has solutions $C_1, C_2, \cdots, C_n$, not all zero. The linear combination

$$y(x) = \sum_{i=1}^{n} C_i y_i(x)$$

of $y_i(x)$ with these coefficients C_i obviously satisfies the equation (8.4) and the initial conditions (8.5) at the point x_0 in D. Therefore, the corollary of Theorem 8.1 implies that

$$y(x) = \sum_{i=1}^{n} C_i y_i(x) \equiv 0$$

This contradicts the fact that $y_1(x), y_2(x), \cdots, y_n(x)$ are linearly independent. Therefore, the Wronskian of linearly independent solutions $y_1(x), y_2(x), \cdots, y_n(x)$ does not vanish at any point in D.

Next we shall consider the Wronskian $W(y_1(x), y_2(x), \cdots, y_n(x))$ of n solutions $y_1(x), y_2(x), \cdots, y_n(x)$ of (8.4) where $y_1(x), y_2(x), \cdots, y_n(x)$ are not necessarily linearly independent. Differentiating $W(y_1(x), y_2(x), \cdots, y_n(x))$ with respect to x, we obtain

$$(9.5) \qquad \frac{d W(y_1(x), y_2(x), \cdots, y_n(x))}{dx}$$

$$= \begin{vmatrix} y_1(x), & y_2(x), & \cdots, & y_n(x) \\ y_1'(x), & y_2'(x), & \cdots, & y_n'(x) \\ \cdots\cdots\cdots\cdots\cdots\cdots\cdots\cdots\cdots \\ y_1^{(n-2)}(x), & y_2^{(n-2)}(x), & \cdots, & y_n^{(n-2)}(x) \\ y_1^{(n)}(x), & y_2^{(n)}(x), & \cdots, & y_n^{(n)}(x) \end{vmatrix}$$

Since $y_i(x)$ satisfies the equation (8.4),

$$y_i^{(n)}(x) = -\sum_{k=1}^{n-1} p_k(x) y_i^{(n-k)}(x) - p_n(x) y_i(x)$$

Substituting this in the above determinant, we obtain

$$(9.6) \quad \frac{dW(y_1(x), y_2(x), \cdots, y_n(x))}{dx}$$
$$= -p_1(x) W(y_1(x), y_2(x), \cdots, y_n(x))$$

Accordingly, $W(y_1(x), y_2(x), \cdots, y_n(x))$ transpose is a solution of the linear homogeneous equation (9.6) with coefficients continuous in D. Therefore, if $W(y_1(x), y_2(x), \cdots, y_n(x))$ vanishes at a point in D, then, by the corollary of Theorem 8.1, $W(y_1(x), y_2(x), \cdots, y_n(x))$ is identically zero in the whole domain D. This proves the following theorem.

THEOREM 9.1. Either the Wronskian of n solutions of (8.4) is identically zero or it never vanishes at any point in D.

By integration of the equation (9.6), we obtain

$$(9.7) \quad W(y_1(x), y_2(x), \cdots, y_n(x))$$
$$= W(y_1(x_0), y_2(x_0), \cdots, y_n(x_0)) \exp\left(-\int_{x_0}^{x} p_1(t)\,dt\right), \quad x \in D$$

which is called *Liouville's formula*. From (8.8), it follows immediately that, if n solusions $y_1(x), y_2(x), \cdots, y_n(x)$ of (8.4) are linearly dependent, then the Wronskian $W(y_1(x), y_2(x), \cdots, y_n(x))$ is identically zero on D. Thus we obtain the following

THEOREM 9.2. Let $y_1(x), y_2(x), \cdots, y_n(x)$ be n solutions of the equation (8.4). Then these solutions are linearly independent if and only if the Wronskian $W(y_1(x), y_2(x), \cdots, y_n(x))$ does not vanish at any point in D. Further, these solutions are linearly dependent if and only if their Wronskian is identically zero in D.

10. *Lagrange's method of variation of constants and D'Alembert's method of reduction of order*

We shall be concerned with the inhomogeneous linear differential equation (8.2). Let $y_1(x), y_2(x)$ be solutions of (8.2). Then, clearly, $y(x) = y_1(x) - y_2(x)$ is a solution of the associated homogeneous equation (8.4). This proves the following theorem.

THEOREM 10.1. The general solution of (8.2) is written as the sum of a particular solution of (8.2) and the general solution of (8.4).

However, if we know a fundamental system of the solutions of (8.4), then we can obtain a particular solution of (8.2) by the *method of variation of constants* which is due to Lagrange. Accordingly, in order to solve linear differential equations, it is sufficient to solve the associated homogeneous equations.

The method of variation of constants. Let $y_1, y_2, \cdots, y_n$ be a fundamental system of the solutions of (8.4). Then the general solution of (8.4) is written in the form

$$(10.1) \qquad\qquad y(x) = \sum_{i=1}^{n} C_i y_n(x)$$

Now we regard these constants C_i as functions of x, and try to determine them in such a way that

$$y(x) = \sum_{i=1}^{n} C_i(x) y_i(x)$$

satisfies (8.2). As was shown by Lagrange, if $C_1(x), C_2(x), \cdots, C_n(x)$ satisfy the system of linear equations

$$y_1(x)C_1'(x) + y_2(x)C_2'(x) + \cdots + y_n(x)C_n'(x) = 0$$
$$y_1'(x)C_1'(x) + y_2'(x)C_2'(x) + \cdots + y_n'(x)C_n'(x) = 0$$
$$(10.2) \qquad \cdots\cdots\cdots\cdots\cdots\cdots\cdots\cdots$$
$$y_1^{(n-2)}(x)C_1'(x) + y_2^{(n-2)}(x)C_2'(x) + \cdots + y_n^{(n-2)}C_n'(x) = 0$$
$$y_1^{(n-1)}(x)C_1'(x) + y_2^{(n-1)}(x)C_2'(x) + \cdots + y_n^{(n-1)}C_n'(x) = q(x)$$

then $\sum_{i=1}^{n} C_i(x) y_i(x)$ satisfies (8.2).

In fact, if there exist $C_1(x), C_2(x), \cdots, C_n(x)$ satisfying (10.2), then, by differentiation and by making use of (10.2), we obtain successively

$$y(x) = \sum_{i=1}^{n} C_i(x) y_i(x)$$

$$y'(x) = \sum_{i=1}^{n} C_i(x) y_i'(x)$$

$$\cdots\cdots\cdots\cdots\cdots$$

$$y^{(n-1)}(x) = \sum_{i=1}^{n} C_i(x) y_i^{(n-1)}(x)$$

$$y^{(n)}(x) = \sum_{i=1}^{n} C_i(x) y_i^{(n)}(x) + q(x)$$

Since $y_i(x)$ satisfies (8.4), $y(x)$ is certainly a solution of (8.2).

Now we consider the system (10.2). According to Theorem 9.2, the Wronskian $W(y_1(x), y_2(x), \cdots, y_n(x))$ of the fundamental system $\{y_i(x)\}$ never vanishes at any point in the domain D, in which the coefficients $p_1(x), p_2(x), \cdots, p_n(x)$ of (8.4) are continuous. Therefore, there exists one and only one set of solutions $C_1'(x), C_2'(x), \cdots, C_n'(x)$ of (10.2), which is written as

$$(10.3) \qquad dC_i(x)/dx = q(x) W_i(x)/ W(y_1(x), y_2(x), \cdots, y_n(x))$$
$$= Z_i(x) , \qquad (i = 1, 2, \cdots, n)$$

where $W_i(x)$ is the cofactor of $y_i^{(n-1)}(x)$ in $W(y_1(x), y_2(x), \cdots, y_n(x))$. Integrating (10.3), we obtain

$$(10.4) \qquad C_i(x) = \int_{x_0}^{x} Z_i(t) \, dt + \bar{C}_i , \qquad (i = 1, 2, \cdots, n)$$

where $\bar{C}_i$ is a constant of integration. Consequently, a particular solution of the equation (8.2) is

$$(10.5) \qquad y(x) = \sum_{i=1}^{n} \left(\int_{x_0}^{x} Z_i(t) \, dt + \bar{C}_i \right) y_i(x)$$

The method of reduction of order. If a particular solution $y_1(x)$, not identically zero, of the nth order linear differential equation (8.4) is known, then, by setting

$$y = y_1 z$$

(8.4) can be reduced to a linear differential equation of the $(n-1)$st order with respect to dz/dx. This procedure is called the *method of reduction of order* and is due to D'Alembert.

In fact, Leibnitz's formula yields

$$y^{(p)} = y_1 z^{(p)} + p y_1' z^{(p-1)} + \cdots + y_1^{(p)} z \qquad (p = 1, 2, \cdots, n)$$

Substituting these in (8.4), we see that the coefficient of $z^{(n)}$ is y_1, and that of z is zero. Thus (8.4) becomes an equation of the $(n-1)$st order with respect to z',

$$(10.6) \qquad y_1 z^{(n)} + q_1(x) z^{(n-1)} + q_2(x) z^{(n-2)} + \cdots + q_{n-1}(x) z' = 0$$

In particular, when $n = 2$, the reduced equation (10.6) can be solved, as was shown by the remark in Part 5. Hence, by virtue of this method, we obtain the general solution

$$(10.7) \qquad y(x) = y_1(x) \int^{x} y_1(t)^{-2} \exp \left(- \int^{t} p_1(\tau) \, d\tau \right) dt$$

$y_1(x)$ being a particular solution of (8.4) with $n = 2$. This method is useful in the practical treatment of the linear differential equations as well as in theoretical studies.

11. *Linear differential equations with constant coefficients*

We consider a homogeneous linear differential equation with

constant coefficients

$$(11.1) \qquad y^{(n)} + a_1 y^{(n-1)} + \cdots + a_{n-1} y' + a_n y = 0$$

For a suitably chosen r, the exponential function

$$(11.2) \qquad y(x) = \exp (rx)$$

satisfies (11.1). To prove this, differentiate (11.2) n times with respect to x. Substituting

$$y' = r \exp (rx), \quad y'' = r^2 \exp (rx), \cdots, \quad y^{(n)} = r^n \exp (rx)$$

in (11.1), we obtain

$$(11.3) \qquad \exp (rx) f(r) = 0$$

where

$$f(r) = r^n + a_1 r^{n-1} + \cdots + a_{n-1} r + a_n$$

Accordingly, if r is a root of the equation

$$(11.4) \qquad f(r) = 0$$

which is called the *characteristic equation* for (11.1), then the function (11.2) satisfies (11.1).

In the case when the characteristic equation (11.4) has n distinct roots $r_1, r_2, \cdots, r_n$, a fundamental system of the solutions of (11.1) is given by

$$(11.5) \qquad e^{r_i x} = \exp (r_i x), \qquad i = 1, 2, \cdots, n$$

In general, the characteristic equation (11.4) has m distinct roots $r_1, r_2, \cdots, r_m$ with multiplicity $\mu_1, \mu_2, \cdots, \mu_m$, such that

$$(11.6) \qquad 1 \leqq m \leqq n, \qquad \sum_{i=1}^{m} \mu_i = n$$

and

$$(11.7) \qquad f(r_i) = 0, \quad f'(r_i) = 0, \cdots, \quad f^{(\mu_i-1)}(r_i) = 0 \quad (i = 1, 2, \cdots, m)$$

In this case, a fundamental system of the solutions of (11.1) is given by

$$(11.8) \qquad \begin{aligned} & e^{r_1 x}, xe^{r_1 x}, \cdots, x^{\mu_1-1} e^{r_1 x} \\ & e^{r_2 x}, xe^{r_2 x}, \cdots, x^{\mu_2-1} e^{r_2 x} \\ & \cdots\cdots\cdots\cdots\cdots\cdots\cdots \\ & e^{r_m x}, xe^{r_m x}, \cdots, x^{\mu_m-1} e^{r_m x} \end{aligned}$$

Proof. We shall first prove that each function in (11.8) satisfies the equation (11.1). To prove this, differentiate with respect to r the identity

$$(e^{rz})^{(n)} + a_1(e^{rz})^{(n-1)} + \cdots + a_n e^{rz} = e^{rz} f(r)$$

Then we have

$$(xe^{rz})^{(n)} + a_1(xe^{rz})^{(n-1)} + \cdots + a_n(xe^{rz}) = e^{rz}(f'(r) + xf(r))$$

Repeating differentiation, we obtain successively

$$(x^2 e^{rz})^{(n)} + a_1(x^2 e^{rz})^{(n-1)} + \cdots + a_n(x^2 e^{rz})$$
$$= e^{rz}(f''(r) + 2xf'(r) + x^2 f(r))$$
$$\cdots\cdots\cdots\cdots\cdots\cdots\cdots$$
$$(x^k e^{rz})^{(n)} + a_1(x^k e^{rz})^{(n-1)} + \cdots + a_n(x^k e^{rz})$$
$$= e^{rz}(f^{(k)}(r) + kxf^{(k-1)}(r) + \tfrac{1}{2}k(k-1)x^2 f^{(k-2)}(r) + \cdots + x^k f(r))$$

Then from (11.7) it follows immediately that each function in (11.8) satisfies (11.1).

Next we shall prove that the functions in (11.8) form a fundamental system. Suppose that the functions in (11.8) are linearly dependent. Since a linear combination of these functions with coefficients $C_1, C_2, \cdots, C_n$ is written in the form

$$P_1(x) e^{r_1 x} + P_2(x) e^{r_2 x} + \cdots + P_m(x) e^{r_m x}$$

where

$$P_i(x) = B_i x^{\nu_i} + \cdots = \text{polynomial of degree } \nu_i \leqq \mu_i - 1$$

we may assume without loss of generality that $P_1(x)$ is not identically zero and

(11.9) $$P_1(x) + P_2(x) e^{(r_2-r_1)x} + \cdots + P_m(x) e^{(r_m-r_1)x} \equiv 0$$

Differentiating this with respect to x, we obtain

$$P_1'(x) + [P_2'(x) + (r_2 - r_1)P_2(x)] e^{(r_2-r_1)x} + \cdots \equiv 0$$

Here $P_1'(x)$ is a polynomial of degree $\nu_1 - 1$, and $[P_2'(x) + (r_2 - r_1)P_2(x)]$ is of degree ν_2 and not identically zero whenever $P_2(x)$ is not identically zero. Accordingly, repeating differentiation ν_1 times, we obtain finally

(11.10) $$Q_2(x) e^{(r_2-r_1)x} + \cdots + Q_m(x) e^{(r_m-r_1)x} \equiv 0$$

where $Q_i(x)$ is a polynomial of degree ν_i whenever $P_i(x)$ is so. On the other hand, by the assumption, at least one of $P_i(x)$, $i = 2, 3, \cdots, m$ is not identically zero, for, if otherwise, $P_1(x)$ must be identically zero, which contradicts the assumption. Therefore, we see that at least one of $Q_i(x)$, $i = 2, 3, \cdots, m$ is not identically zero. Repeating the same argument, we finally obtain that there exists a polynomial $V(x)$ such that

$$V(x)e^{sx} \equiv 0 , \qquad V(x) \not\equiv 0$$

This is a contradiction. Hence the linear independence of the functions in (11.8) is proved. Since the equation (11.1) is of order n and the number of the functions in (11.8) is also n, the system (11.8) is a fundamental system, q.e.d.

REMARK 1. We consider the case when the coefficients $a_1, a_2, \cdots, a_n$ of (11.1) are all real. In this case, if the characteristic equation (11.4) has a complex root

$$r_1 = \alpha + i\beta \qquad (i = \sqrt{-1})$$

then its conjugate

$$r_2 = \alpha - i\beta$$

is also a root of (11.4) with the same multiplicity μ. The corresponding solution to these roots r_1, r_2 is written in the form

$$(11.11) \qquad P_1(x)e^{r_1 x} + P_2(x)e^{r_2 x}$$

where $P_1(x)$, $P_2(x)$ are polynomials of degree $\mu - 1$. Put

$$R_1(x) = P_1(x) + P_2(x), \quad R_2(x) = i\{P_1(x) - P_2(x)\}$$

Then (11.11) becomes

$$(11.12) \qquad e^{\alpha x}\{R_1(x) \cos \beta x + R_2(x) \sin \beta x\}$$

Accordingly, if we are concerned only with real solutions, we may take $R_1(x), R_2(x)$ in (11.12) as polynomials with real coefficients.

REMARK 2. We can obtain a particular solution of the inhomogeneous equation

$$(11.13) \qquad y^{(n)} + a_1 y^{(n-1)} + \cdots + a_{n-1} y' + a_n y = b(x)$$

from a fundamental system of the homogeneous equation (11.1) by the method of variation of constants. However, if $b(x)$ is of the form

$$(11.14) \qquad b(x) = e^{kx}(b_0 x^p + b_1 x^{p-1} + \cdots + b_{p-1} x + b_p)$$

we can find it directly as follows. Namely, we can prove the following: (i) if k is not a root of the characteristic equation (11.4), then there exists a particular solution of (11.13) which is of the form

$$y = e^{kx} \text{ (polynomial of degree } p)$$

(ii) if k is a root of (11.4) with multiplicity μ, then there exists a particular solution of (11.13) which is of the form

$$y = e^{kx} \text{ (polynomial of degree } (p + \mu))$$

Proof. Suppose that a function y of the form

$$y = e^{kx}z$$

satisfies the equation (11.13). The derivatives of y are given by

$$y' = e^{kx}(kz + z')$$
$$y'' = e^{kx}(k^2z + 2kz' + z'')$$
$$\dots\dots\dots\dots\dots\dots$$
$$y^{(n)} = e^{kx}(k^nz + nk^{n-1}z' + \cdots + z^{(n)})$$

Hence, substituting these in (11.13), we obtain

$$(11.15) \qquad f(k)z + \frac{f'(k)}{1!}z' + \frac{f''(k)}{2!}z'' + \cdots + \frac{f^{(n)}(k)}{n!}z^{(n)}$$
$$= b(x)e^{-kx}, \quad f(k) = k^n + a_1k^{n-1} + \cdots + a_{n-1}k + a_n$$

which is a sufficient condition that the function y satisfies (11.13).

In case (i), it is easily seen that a function z of the form

$$z = B_0x^p + B_1x^{p-1} + \cdots + B_{p-1}x + B_p$$

satisfies (11.15). In fact, since $f(k) \neq 0$, we can determine the coefficients B_i successively as follows:

$$f(k)B_0 = b_0, \quad f(k)B_1 + pf'(k)B_0 = b_1, \cdots$$

In case (ii) we have

$$f(k) = 0, f'(k) = 0, f''(k) = 0, \cdots, f^{(\mu-1)}(k) = 0, f^{(\mu)}(k) \neq 0$$

Hence (11.15) becomes

$$(11.16) \qquad \frac{f^{(\mu)}(k)}{\mu!}z^{(\mu)} + \frac{f^{(\mu+1)}(k)}{(\mu+1)!}z^{(\mu+1)} + \cdots + \frac{f^{(n)}(k)}{n!}z^{(n)} = b(x)e^{-kx}$$

Since $f^{(\mu)}(k) \neq 0$, setting

$$z^{(\mu)} = B_0x^p + B_1x^{p-1} + \cdots + B_{p-1}x + B_p$$

we can determine the coefficients B_i by (11.16) as follows

$$\frac{f^{(\mu)}(k)}{\mu!} B_0 = b_0$$

$$\frac{f^{(\mu)}(k)}{\mu!} B_1 + \frac{f^{(\mu+1)}(k)}{(\mu+1)!} p B_0 = b_1$$

. . . .

EXAMPLE 1. $y'' - y = e^{2x}$

The associated characteristic equation has two roots 1 and -1. Hence the equation possesses a solution of the form Ce^{2x}. To determine the constant C, substitute Ce^{2x} for y in the equation. Then we obtain $C = \frac{1}{3}$. Therefore, the general solution is given by

$$y = C_1 e^x + C_2 e^{-x} + \tfrac{1}{3} e^{2x}$$

EXAMPLE 2. $y'' + y = \sin 2x$

The associated characteristic equation is $r^2 + r = 0$, and has two roots $\pm i$. To apply the above method, consider inhomogeneous equations

$$y_1'' + y_1 = \frac{1}{2i} e^{2ix}, \qquad y_2'' + y_2 = \frac{-1}{2i} e^{-2ix}$$

Let y_1 and y_2 be particular solutions of the above equations respectively. Then, because of

$$\sin 2x = \frac{1}{2i} e^{2ix} - \frac{1}{2i} e^{-2ix}$$

$y_1 + y_2$ is a solution of the original equations. On the other hand, both $\pm 2i$ are not roots of the characteristic equation. Hence we seek for the particular solutions

$$y_1 = C e^{2ix}, \qquad y_2 = E e^{-2ix}$$

of the above equations and obtaining, respectively, $C = -E = -1/6i$. Thus the original equation possesses a particular solution $-\frac{1}{3} \sin 2x$. Therefore the general solution is given by

$$y = C_1 \sin x + C_2 \cos x - \tfrac{1}{3} \sin 2x$$

EXERCISE 1. Solve the equations in Examples 1 and 2 by the method of variation of constants.

EXERCISE 2. An equation of the form

(11.17) $x^n y^{(n)} + a_1 x^{n-1} y^{(n-1)} + a_2 x^{n-2} y^{(n-2)} + \cdots + a_n y = 0$

is called the *Euler differential equation*. Show that by the transformation

(11.18) $$x = \exp t$$

the equation (11.17) is transformed into a linear homogeneous differential equation with constant coefficients.

§3. Second order differential equations of the Fuchs type

The theory of functions enables us to study complex solutions of differential equations. In this paragraph, the second order equations of Fuchs type will be dealt with. Our main concern is Fuchs' theorem. This theorem enables us to undertake a unified treatment of various kinds of equations, namely, Gauss equations, Legendre equations and Bessel equations, which are important in practice. In Part 13–15, we shall be concerned with these equations, which will be needed in Chapter 5.

12. *Regular singular points. Fuchs' theorem*

We consider a second order differential equation

(12.1) $$y'' + p_1(x)y' + p_2(x)y = 0$$

where $p_1(x)$ and $p_2(x)$ are complex-valued functions with an isolated singularity at a point x_0, but single-valued analytic in a complex domain K:

(12.2) $$0 < |x - x_0| < a', \qquad a' > 0$$

Let x be a point in K. Then, for any initial conditions at this point x, there exists a solution $y(x)$ of this initial value problem for the equation (12.1). Consider a curve connecting x and x_0 in the domain K. Then, as was shown in Part 7, the analytic continuation of $y(x)$ along the curve is possible up to and except for the point x_0. We denote again by $y(x)$ the solution obtained by such an analytic continuation. When, for any solution $y(x)$, there exists a positive number ρ such that

(12.3) $$\lim_{x \to x_0} (x - x_0)^\rho y(x) = 0$$

the point x_0 is called a *regular singular point* for the equation (12.1). Our subject is to prove the following

THEOREM 12.1. (Fuchs) In order that x_0 is a regular singular point, it is necessary and sufficient that $p_1(x)$ has at most a pole of order 1 and $p_2(x)$ at most a pole of order 2 at the point x_0.

As was shown in Part 7, we can perform the analytic continuation of $y(x)$ along any curve in the domain K and obtain an analytic function, which is, in general, not single-valued in the domain K. We shall prove first that there exists a solution $y(x)$ of the equation (12.1) which is of the form

$$(12.4) \qquad y(x) = (x - x_0)^{r_1} \varphi(x)$$

$\varphi(x)$ being a single-valued analytic function in the domain K. To prove this, let $y_1(x), y_2(x)$ be a fundamental system of the solutions in a neighbourhood of the point x. Let C be a closed curve in K starting from the point x and winding around the point x_0 once in positive sense. By analytic continuations of $y_1(x)$ and $y_2(x)$ along this curve C and returning to the point x, we obtain $Y_1(x)$ and $Y_2(x)$, respectively. $Y_1(x)$ and $Y_2(x)$ may differ from $y_1(x)$ and $y_2(x)$ respectively. However, since $p_1(x)$ and $p_2(x)$ are single-valued analytic so that they do not change their values by the analytic continuation, $Y_1(x)$ and $Y_2(x)$ still satisfy the equation (12.1) in a neighbourhood of x. Hence $Y_1(x)$ and $Y_2(x)$ must be expressible as linear combinations of $y_1(x)$ and $y_2(x)$:

$$(12.5) \qquad \begin{aligned} Y_1(x) &= \alpha y_1(x) + \beta y_2(x) \\ Y_2(x) &= \gamma y_1(x) + \delta y_2(x) \end{aligned}$$

Accordingly, a solution of the form

$$cy_1(x) + dy_2(x)$$

where c, d are arbitrary constants—every solution in a neighbourhood of x is written in such a form—changes its value to

$$\begin{aligned} c\, Y_1(x) + d\, Y_2(x) &= c(\alpha y_1(x) + \beta y_2(x)) + d(\gamma y_1(x) + \delta y_2(x)) \\ &= (c\alpha + d\gamma)y_1(x) + (c\beta + d\delta)y_2(x) \end{aligned}$$

by the analytic continuation along the curve C. If c, d satisfy

$$(12.6) \qquad c\alpha + d\gamma = \lambda c, \qquad c\beta + d\delta = \lambda d$$

for some constant λ, then the solution

$$(12.7) \qquad y(x) = cy_1(x) + dy_2(x)$$

changes its value to $\lambda y(x)$ by the analytic continuation along the curve C. Such a factor λ must be an eigenvalue of the matrix

$$\begin{bmatrix} \alpha & \gamma \\ \beta & \delta \end{bmatrix}$$

In other words, λ must be a root of the equation

$$(12.8) \qquad \begin{vmatrix} \alpha - \lambda & \gamma \\ \beta & \delta - \lambda \end{vmatrix} = 0$$

Since $y_1(x)$ and $y_2(x)$ are linearly independent, $Y_1(x)$ and $Y_2(x)$ are also linearly independent. For if otherwise, the analytic continuation of $Y_1(x)$ and $Y_2(x)$ along the curve C in negative sense would yield the linear dependence of $y_1(x)$ and $y_2(x)$. Hence $\alpha\delta - \beta\gamma \neq 0$. This implies that the equation (12.8) possesses a non-zero root λ. For this λ there exists certainly a set of solutions c, d of (12.6) such that $|c| + |d| \neq 0$. Take such c, d and consider the solution $y(x) = cy_1(x) + dy_2(x) \not\equiv 0$. Let

$$(12.9) \qquad \exp(2\pi\sqrt{-1}\, r_1) = \lambda$$

(Such r_1 does exist since λ is not zero.) By the analytic continuation of $(x-x_0)^{r_1} = \exp(r_1 \log(x-x_0))$ along the curve C and returning to the point x, we obtain $\lambda(x - x_0)^{r_1}$. Hence

$$(12.10) \qquad (x - x_0)^{-r_1} y(x) = \varphi(x)$$

does not change its value by this analytic continuation. This means that $\varphi(x)$ is single-valued analytic in the domain K.

Proof of Theorem 12.1. *Necessity.* The equation (12.1) possesses a solution $y_1(x)$ of the form

$$y_1(x) = (x - x_0)^{r_1} \varphi_0(x)$$

as was shown above. Since x_0 is a regular singular point for (12.1), $\varphi_0(x)$ has at most a pole at the point x_0; for if otherwise, the Laurent expansion of $\varphi_0(x)$ at x_0 actually contains infinitely many powers of $(x - x_0)^{-1}$ and hence (12.3) would not hold. Thus $y_1(x)$ may be written as

$$(12.11) \qquad y_1(x) = (x - x_0)^{r} \varphi(x)$$

where $\varphi(x)$ is single-valued analytic in the domain $|x - x_0| < a$ and $\varphi(x_0) \neq 0$. To find another solution, linearly independent of y_1, by the method of reduction of order, set

$$y_2(x) = y_1(x) \int^{x} w(t)\, dt$$

Then $w(x)$ satisfies the equation

$$(12.12) \qquad \frac{dw}{dx} + \left\{ p_1(x) + \frac{2y_1'(x)}{y_1(x)} \right\} w = 0$$

By analytic continuation of $w(x)$ along the curve C starting from the point x, we obtain $W(x)$, when we return to the point x.

Since $p_1(x)$ and $y_1'(x)/y_1(x)$ are single-valued analytic in the domain K so that they do not change their values by this continuation, $W(x)$ is also a solution of the equation (12.12) in a neighbourhood of x. Hence there exists a constant $\eta \neq 0$ such that

$$W(x) = \eta \, w(x)$$

Therefore $w(x)$ can be expressed as

$$w(x) = (x - x_0)^s \psi_1(x), \quad \exp(2\pi\sqrt{(-1)}s) = \eta$$

where $\psi_1(x)$ is a single-valued analytic function in the domain K. Thus $y_2(x)$ can be expressed as

$$(12.13) \qquad y_2(x) = y_1(x)\{h \log(x - x_0) + (x - x_0)^s \psi(x)\}$$

where h is a constant, $\psi(x)$ is a single-valued analytic function in the domain K. Further, by virtue of (12.3), $\psi(x)$ has at most a pole at the point x_0. The logarithmic term $h \log(x - x_0)$ appeares in (12.13) only when the Laurent expansion of $\psi_1(x)$ at the point x_0 contains the term $(x - x_0)^{-s-1}$.

Since $y_1(x)$ and $y_2(x)$ are linearly independent solutions of (12.1), $p_1(x)$ is expressed as

$$
\begin{aligned}
p_1(x) &= \frac{y_1''(x)\,y_2(x) - y_2''(x)\,y_1(x)}{y_1(x)\,y_2'(x) - y_2(x)\,y_1'(x)} \\
&= -\frac{d}{dx}\left\{\log\left[y_1(x)^2 \frac{d}{dx}\left(\frac{y_2(x)}{y_1(x)}\right)\right]\right\} \\
&= -\frac{2y_1'(x)}{y_1(x)} - \frac{d}{dx}\left\{\log\frac{d}{dx}\left(\frac{y_2(x)}{y_1(x)}\right)\right\}
\end{aligned}
$$

as was shown in Part 9. Hence, by virtue of (12.11) and (12.13), $p_1(x)$ has at most a pole of order 1 at the point x_0. Since $p_2(x)$ is given by

$$p_2(x) = -\frac{y_1''(x)}{y_1(x)} - p_1(x)\frac{y_1'(x)}{y_1(x)}$$

the above result, together with (12.11), implies that $p_2(x)$ has at most a pole of order 2 at the point x_0.

Sufficiency. According to the assumption, the equation (12.1) can be written as

$$(12.14) \qquad (x - x_0)^2 y'' + (x - x_0) P(x) y' + Q(x) y = 0$$

where $P(x)$ and $Q(x)$ are single-valued analytic in the domain $|x - x_0| < a'$. Let the Taylor expansions of $P(x)$ and $Q(x)$ be

(12.15)
$$P(x) = p_0 + p_1(x - x_0) + p_2(x - x_0)^2 + \cdots$$
$$Q(x) = q_0 + q_1(x - x_0) + q_2(x - x_0)^2 + \cdots$$

respectively. Let a function $y(x)$ of the form

(12.16)
$$y(x) = (x - x_0)^r \left\{ 1 + \sum_{n=1}^{\infty} a_n(x - x_0)^n \right\}$$

satisfy the equation (12.14). Substituting (12.16) in (12.14), we obtain

$$(x - x_0)^r \left\{ r(r - 1) + \sum_{n=1}^{\infty} a_n(r + n)(r + n - 1)(x - x_0)^n \right\}$$
$$+ (x - x_0)^r P(x) \left\{ r + \sum_{n=1}^{\infty} a_n(r + n)(x - x_0)^n \right\}$$
$$+ (x - x_0)^r Q(x) \left\{ 1 + \sum_{n=1}^{\infty} a_n(x - x_0)^n \right\} = 0$$

Then, substituting (12.15) in this relation, we obtain, by a comparison of coefficients, the following relations:

$$r^2 + (p_0 - 1)r + q_0 = 0$$
$$a_1\{(r + 1)^2 + (p_0 - 1)(r + 1) + q_0\} + rp_1 + q_1 = 0$$
$$a_2\{(r + 2)^2 + (p_0 - 1)(r + 2) + q_0\}$$
$$\qquad + a_1\{(r + 1)p_1 + q_1\} + rp_2 + q_2 = 0$$

(12.17)
$$\cdots\cdots\cdots\cdots\cdots\cdots\cdots\cdots$$

$$a_n\{(r + n)^2 + (p_0 - 1)(r + n) + q_0\}$$
$$\qquad + \sum_{m=1}^{n-1} a_{n-m}\{(r + n - m)p_m + q_m\} + rp_n + q_n = 0$$

$$\cdots\cdots\cdots\cdots\cdots\cdots\cdots\cdots$$

Hence r must satisfy the equation

(12.18)
$$F(r) = r^2 + (p_0 - 1)r + q_0 = 0$$

This equation is called the *indicial equation* for (12.14) at the point x_0. Let r_1 and r_2 be two roots of the indicial equation (12.18) and let

(12.19)
$$s = r_1 - r_2, \quad \text{real part of } s \geqq 0$$

Then, by making use of $F(r_1) = 0$ and $(p_0 - 1) + 2r_1 = s$, we obtain

$$F(r_1 + n) = (r_1 + n)^2 + (p_0 - 1)(r_1 + n) + q_0$$
$$= F(r_1) + 2nr_1 + n^2 + (p_0 - 1)n$$
$$= n^2 + n\{(p_0 - 1) + 2r_1\} = n(s + n)$$

namely,

$$(12.20) \qquad\qquad F(r_1 + n) = n(s + n)$$

Because of (12.19), $F(r_1 + n) \neq 0$ for any $n = 1, 2, 3, \cdots$. Hence, by setting $r = r_1$, the coefficients a_n are determined successively by the relations (12.17). If the resulting power series

$$(12.21) \qquad\qquad \sum_{n=1}^{\infty} a_n(x - x_0)^n$$

has a positive radius of convergence, then the equation (12.14), and hence (12.1), possess a genuine solution $y_1(x)$ of the form

$$(12.16') \qquad y_1(x) = (x - x_0)^{r_1}\left\{1 + \sum_{n=1}^{\infty} a_n(x - x_0)^n\right\}$$

Moreover, the series (12.16') converges in the whole domain K. For, as was shown in Part 7, the solution $y_1(x)$ has no singular points except for the point x_0, since $P(x)$ and $Q(x)$ are single-valued analytic in the domain $|x - x_0| < a'$.

We shall prove that the radius of convergence of (12.21) is positive. Since the two series in (12.15) converge in the domain $|x - x_0| < a$, for any $a < a'$, there exists a positive number M such that

$$(12.22) \qquad |p_n| < Ma^{-n}, \quad |q_n| < Ma^{-n} \qquad (n = 0, 1, 2, \cdots)$$

Hence we may assume without loss of generality that

$$(12.23) \qquad |r_1 p_n + q_n| \leq Ma^{-n}, \quad M > 1 \qquad (n = 0, 1, 2, \cdots)$$

Since $|s + 1| \geq 1$ by (12.19), (12.20) yields

$$|a_1| = |(r_1 p_1 + q_1)/F(r_1 + 1)| \leq M/a(|s + 1|) \leq M/a$$

In general, we can prove by induction that

$$(12.24) \qquad\qquad |a_n| \leq M^n a^{-n} \qquad (n = 1, 2, 3, \cdots)$$

In fact, suppose that (12.24) holds for $n = 1, 2, 3, \cdots, m - 1$. Then, by making use of the fact that $M > 1$ and $|1 + sm^{-1}| \geq 1$, we obtain

$$|a_m| = |F(r_1 + m)|^{-1} \left| \sum_{t=1}^{m-1} a_{m-t}\{(r_1 + m - t)p_t + q_t\} + r_1 p_m + q_m \right|$$

$$\leq (m|s + m|)^{-1} \left\{ \sum_{t=1}^{m-1} |a_{m-t}| |r_1 p_t + q_t| + |r_1 p_m + q_m| \right.$$

$$\left. + \sum_{t=1}^{m-1} (m - t)|a_{m-t}| |p_t| \right\}$$

$$\leq (m^2|1 + sm^{-1}|)^{-1} \left\{ mM^m a^{-m} + M^m a^{-m} \sum_{t=1}^{m-1} (m - t) \right\}$$

$$\leq m^{-2}\left(m + \frac{m(m-1)}{2} \right) M^m a^{-m}$$

$$= \frac{m + 1}{2m} M^m a^{-m} < M^m a^{-m}$$

Therefore, (12.24) holds for every $n = 1, 2, 3, \cdots$. This implies that the radius of convergence of (12.21) is positive.

We thus obtain a solution $y_1(x)$, not identically zero, which satisfies (12.3). Next, we shall find another solution which is linearly independent of $y_1(x)$. We shall consider first the case when

(12.25) $\qquad s = r_1 - r_2 \neq$ positive integer

From (12.25) and

$$F(r_2 + n) = n(-s + n)$$

it follows that

$$\frac{1}{F(r_2 + n)} = \frac{1}{n^2\left(1 - \dfrac{s}{n}\right)} \leq \kappa/n^2$$

where

$$\kappa = \sup_{n \geq 1} \left| 1 - \frac{s}{n} \right|^{-1}$$

Hence we can prove similarly as before that there exists a solution $y_2(x)$ of the form

(12.26) $\qquad y_2(x) = (x - x_0)^{r_2}\left\{ 1 + \sum_{n=1}^{\infty} b_n(x - x_0)^n \right\}$

which is linearly independent of $y_1(x)$. That the radius of convergence of the series in the brace on the right side of (12.26) is positive follows from

$$(12.27) \qquad |b_n| \leq (M\kappa)^n a^{-n} \qquad (n = 1, 2, 3, \cdots)$$

which can be proved similarly as before.

Next, we shall consider the case when (12.25) does not hold. By the method of reduction of order, that is, by setting

$$(12.28) \qquad y = y_1 \int^x z(t)\, dt$$

we obtain

$$(12.29) \qquad z' + \left(\frac{P(x)}{(x - x_0)} + \frac{2y_1'(x)}{y_1(x)} \right) z = 0$$

Since $P(x)/(x - x_0) + 2y_1'(x)/y_1(x)$ has at most a pole of order 1 at the point x_0, the solution $z(x)$ of (12.29) is given by

$$z(x) = \exp \left(\int^x - \left\{ \frac{P(t)}{(t - x_0)} + \frac{2y_1'(t)}{y_1(t)} \right\} dt \right)$$
$$= (x - x_0)^\alpha \{ \text{convergent power series in } (x - x_0) \}$$

Hence there exists a solution $y_2(x)$, linearly independent of $y_1(x)$, which is of the form

$$y_2(x) = y_1(x) \int^x z(t)\, dt$$
$$= y_1(x) \{ \beta \log (x - x_0) + \varphi_2(x)(x - x_0)^\alpha \}$$

Here $\varphi_2(x)$ is a single-valued analytic function in the domain $|x - x_0| < a'$.

Thus the existence of the fundamental system of the solutions $y_1(x)$ and $y_2(x)$ each of which satisfies (12.3) is proved, q.e.d.

REMARK. The solutions need not have singularities at the regular singular points. In the following example, it is shown that every solution is analytic at the regular singular point $x = x_0$. Consider the equation

$$y'' - \frac{2}{(x - x_0)} y' + \frac{2}{(x - x_0)^2} y = 0$$

The point $x = x_0$ is a regular singular point for the equation, however, a fundamental system of the solutions is given by

$$y_1 = (x - x_0), \qquad y_2 = (x - x_0)^2$$

each of which is analytic at the point $x = x_0$.

Singularities at infinity. By the transformation of variable,

$$x = t^{-1}$$

the equation (12.1),

$$y'' + p_1(x)y' + p_2(x)y = 0$$

is transformed into

(12.30) $$\frac{d^2y}{dt^2} + \left\{ \frac{2}{t} - \frac{p_1(t^{-1})}{t^2} \right\} \frac{dy}{dt} + \frac{p_2(t^{-1})}{t^4} y = 0$$

The point $x = \infty$ is said to be a regular singular point for the equation (12.1) when the point $t = 0$ is a regular singular point for the equation (12.30), in other words, when

(12.31)
$$p_1(x) = O(x^{-1}) = p_0 x^{-1} + O(x^{-2})$$
$$p_2(x) = O(x^{-2}) = q_0 x^{-2} + O(x^{-3})$$

as $x \to \infty$. The indicial equation at the point $x = \infty$ is that for the equation (12.30) at the point $t = 0$, hence it is given by

(12.32) $$r^2 + (1 - p_0)r + q_0 = 0$$

13. Gauss differential equations

The *Gauss differential equation* (or *hypergeometric differential equation*) is the equation of the form

(13.1) $$x(1 - x)y'' + \{\gamma - (\alpha + \beta + 1)x\}y' - \alpha\beta y = 0$$

where α, β and γ are constants. Making the transformation

(13.2) $$x = 1 - \xi$$

in (13.1), then replacing ξ by x, we obtain the equation

(13.3) $$x(1 - x)y'' + \{\alpha + \beta - \gamma + 1 - (\alpha + \beta + 1)x\}y' - \alpha\beta y = 0$$

Making the transformation

(13.4) $$x = \xi^{-1}, \qquad y = \xi^{-\alpha}\eta$$

in (13.1), then replacing ξ by x and η by y, we obtain the equation

(13.5) $$x(1 - x)y'' + \{\alpha - \beta + 1 - (2\alpha - \gamma + 2)x\}y'$$
$$- \alpha(\alpha - \gamma + 1)y = 0$$

Making the transformation

(13.6) $$y = x^{1-\gamma}\eta$$

in (13.1), then replacing η by y, we obtain the equation

$$(13.7) \qquad x(1 - x)y'' + \{2 - \gamma - (\alpha + \beta - 2\gamma + 3)x\}y'$$
$$- (\alpha - \gamma + 1)(\beta - \gamma + 1)y = 0$$

(These calculations will be left for the reader.)

Obviously, the point $x = 0$ is a regular singular point for the equation (13.1). The indicial equation at this point is

$$(13.8) \qquad\qquad r^2 + (\gamma - 1)r = 0$$

and has roots 0 and $(1 - \gamma)$. According to (13.2) and (13.3), the point $x = 1$ is a regular singular point for the equation (13.1). The roots of the indicial equation at this point are 0 and $1 - (\alpha + \beta - \gamma + 1) = \gamma - \alpha - \beta$. Further, according to (13.4) and (13.5), the point $x = \infty$ is also a regular singular point for the equation and the roots of the indicial equation at this point are $0 + \alpha = \alpha$ and $1 - (\alpha - \beta + 1) + \alpha = \beta$.

It is easily seen that the equation (13.1) has no singular points except for the points 0, 1, and ∞. Accordingly, all the singular points for the Gauss equation are regular singular points. Such an equation is said to be of the *Fuchs type*. In order to show this property and to show that the roots of the corresponding indicial equations are

$$0, 1 - \gamma; \quad 0, \gamma - \alpha - \beta; \quad \alpha, \beta$$

we denote the general solution $y(x)$ of the equation (13.1) by

$$(13.9) \qquad y(x) = P \left\{ \begin{matrix} 0 & 1 & \infty & \\ 0 & 0 & \alpha & x \\ 1 - \gamma & \gamma - \alpha - \beta & \beta & \end{matrix} \right\}$$

which is called *Riemann's notation*.

In the following, we shall restrict ourselves to the case for which

$$(13.10) \qquad \text{none of } \gamma, \ \gamma - \alpha - \beta, \ \alpha - \beta \text{ is an integer.}$$

To begin with, we consider the fundamental system of the solutions at the point $x = 0$. Since the indicial equation at this point has the root 0, there exists a solution $y_1(x)$ of the form

$$y_1(x) = 1 + \sum_{n=1}^{\infty} a_n x^n$$

which converges for $|x| < 1$. Substituting this in (13.1), we obtain, by comparison of coefficients, the following *recurrence relations*

$$a_n\{n(n - 1 + \gamma)\} - a_{n-1}\{(n - 1 + \alpha)(n - 1 + \beta)\} = 0$$

Since $a_0 = 1$, we obtain

$$(13.11) \quad a_n = \frac{\alpha(\alpha + 1) \cdots (\alpha + n - 1) \beta(\beta + 1) \cdots (\beta + n - 1)}{1 \cdot 2 \cdots n \cdot \gamma(\gamma + 1) \cdots (\gamma + n - 1)}$$

Hence, the *hypergeometric series*

$$(13.12) \qquad y_{10}(x) = F(\alpha, \beta, \gamma, x)$$

$$= 1 + \frac{\alpha \cdot \beta}{1 \cdot \gamma} x + \frac{\alpha(\alpha + 1) \beta(\beta + 1)}{1 \cdot 2 \cdot \gamma(\gamma + 1)} x^2 + \cdots$$

converges for $|x| < 1$ and satisfies the equation (13.1) under the assumption (13.10). (It is known that, if $\gamma \neq 0, -1, -2, \cdots$, the series (13.12) is convergent.) By virtue of the transformation (13.6) and the transformed equation (13.7), it is easily seen that another solution $y_{20}(x)$ which is linearly independent of $y_{10}(x)$ is of the form

$$(13.13) \quad . \qquad y_{20}(x) = x^{1-\gamma} F(\alpha - \gamma + 1, \beta - \gamma + 1, 2 - \gamma, x)$$

Similarly, by virtue of the transformation (13.2) and the transformed equation (13.3), the fundamental system of the solutions at the point $x = 1$ is given by

$$(13.14) \quad \begin{aligned} y_{11}(x) &= F(\alpha, \beta, \alpha + \beta - \gamma + 1, 1 - x) \\ y_{21}(x) &= (1 - x)^{\gamma - \alpha - \beta} F(\gamma - \beta, \gamma - \alpha, \gamma - \alpha - \beta + 1, 1 - x) \end{aligned}$$

each of which converges for $|x - 1| < 1$.

Further, by virtue of the transformation (13.4) and the transformed equation (13.5), the fundamental system of the solutions at the point $x = \infty$ is given by

$$(13.15) \quad \begin{aligned} y_{1\infty}(x) &= x^{-\alpha} F(\alpha, 1 + \alpha - \gamma, 1 + \alpha - \beta, x^{-1}) \\ y_{2\infty}(x) &= x^{-\beta} F(\beta, 1 + \beta - \gamma, 1 + \beta - \alpha, x^{-1}) \end{aligned}$$

each of which converges for $|x^{-1}| < 1$.

REMARK. The hypergeometric series includes many kinds of Taylor series of elementary functions as special cases. For example,

$$(1 - x)^n = F(-n, 1, 1, x)$$

$$\log (1 - x) = xF(1, 1, 2, x)$$

$$\exp x = \lim_{\beta \to \infty} F\left(1, \beta, 1, \frac{x}{\beta}\right)$$

$$\sin x = \lim_{\alpha, \beta \to \infty} F\left(\alpha, \beta, \frac{3}{2}, -\frac{x^2}{4\alpha\beta}\right)$$

$$\cos x = \lim_{\alpha, \beta \to \infty} F\left(\alpha, \beta, \frac{1}{2}, -\frac{x^2}{4\alpha\beta}\right)$$

14. *Legendre differential equations*

We consider a Gauss equation with coefficients

(14.1) $\alpha = n + 1$, $\beta = -n$, $\gamma = 1$,   (n = integer)

that is,

(14.2) $x(1 - x)y'' + (1 - 2x)y' + n(n + 1)y = 0$

Changing the variable x by the transformation

(14.3) $x = \dfrac{1 - \xi}{2}$

then denoting the new variable ξ again by x, we obtain the equation

(14.4) $(1 - x^2)y'' - 2xy' + n(n + 1)y = 0$

The equation (14.4) is known as the *Legendre differential equation.*
The equation (14.4) can also be written in the form

(14.5) $\{(1 - x^2)y'\}' + n(n + 1)y = 0$

It is easily seen that the singular points for this equation are $1, -1$,
and ∞, and that these points are regular singular points. Hence
there exists a fundamental system of solutions, each of which is
expressed by a convergent power series for $|x| < 1$. Let

$$y = C_0 + C_1 x + \cdots + C_\nu x^\nu + \cdots$$

Substituting this in (14.4), we obtain, by comparison of coefficients,
the following recurrence relations

$$(\nu + 1)(\nu + 2)C_{\nu+2} + (n - \nu)(n + \nu + 1)C_\nu = 0$$

Hence, setting $C_0 = 1$, $C_1 = 0$, and $C_0 = 0$, $C_1 = 1$, we have the
solutions

(14.6)
$$y_1 = 1 - \frac{n(n + 1)}{2!}x^2 + \frac{n(n + 1)(n - 2)(n + 3)}{4!}x^4 + \cdots$$

$$y_2 = x - \frac{(n - 1)(n + 2)}{3!}x^3$$

$$+ \frac{(n - 1)(n + 2)(n - 3)(n + 4)}{5!}x^5 + \cdots$$

respectively. Obviously, y_1 and y_2 form a fundamental system of
solutions. Further, if n is a positive integer, then either $y_1(x)$ or
$y_2(x)$ becomes a polynomial of degree n, $y_1(x)$ for even n, $y_2(x)$ for
odd n. This polynomial of degree n coincides with the nth *Legendre*

polynomial (or the *n*th *zonal harmonics*)

$$(14.7) \qquad P_n(x) = \frac{(2n)!}{2^n(n!)^2} \left\{ x^n - \frac{n(n-1)}{2(2n-1)} x^{n-2} \right.$$
$$\left. + \frac{n(n-1)(n-2)(n-3)}{2 \cdot 4 \cdot (2n-1)(2n-3)} x^{n-4} - \cdots \right\}$$

except for a multiplicative factor.

The indicial equation at the point $x = \infty$ is

$$r(r+1) - 2r - n(n+1) = 0$$

and has two roots $n+1$ and $-n$. Hence, for any positive integer n, there exists a solution y of the form

$$y = g_0 x^{-(n+1)} + g_1 x^{-(n+2)} + \cdots$$

which converges for $|x| > 1$. Substituting this in (14.4) and setting

$$g_0 = \frac{n!}{3 \cdot 5 \cdot \ \cdots \ (2n+1)}$$

we obtain, by comparison of coefficients, that the solution y coincides with the *Legendre function of the second kind* $Q_n(x)$, which is defined as

$$(14.8) \qquad Q_n(x) = \frac{2^n(n!)^2}{(2n+1)! \, x^{n+1}} \left\{ 1 + \frac{(n+1)(n+2)}{2(2n+3)} \frac{1}{x^2} \right.$$
$$\left. + \frac{(n+1)(n+2)(n+3)(n+4)}{2 \cdot 4 \cdot (2n+3)(2n+5)} \frac{1}{x^4} + \cdots \right\}$$

It is easily seen that $Q_n(x)$ and $P_n(x)$ form a fundamental system of the solutions.

EXERCISE. Show that

$$(14.9) \qquad P_n(x) = F\left(n+1, -n, 1, \frac{1-x}{2} \right)$$

$$(14.10) \quad Q_n(x) = \frac{2^n(n!)^2}{(2n+1)!} \frac{1}{x^{n+1}} F\left(\frac{n+1}{2}, \frac{n+2}{2}, \frac{2n+3}{2}, x^{-2} \right)$$

Generating function. According to the binomial theorem, we see that, for $|2tx - t^2| < 1$,

$$(1 - 2xt + t^2)^{-\frac{1}{2}} = (1 - (2xt - t^2))^{-\frac{1}{2}}$$
$$= \sum_{m=0}^{\infty} \frac{(2m)!}{2^{2m}(m!)^2} (2xt - t^2)^m$$

$$= \sum_{m=0}^{\infty} \frac{(2m)!}{2^{2m}(m!)^2} \sum_{s=0}^{m} (-1)^s \frac{2^{m-s}m!}{s!(m-s)!} x^{m-s} t^{m+s}$$

$$= \sum_{m=0}^{\infty} \sum_{s=0}^{m} (-1)^s \frac{(2m)!}{2^{m+s}m! \, s!(m-s)!} x^{m-s} t^{m+s}$$

Hence we obtain

$$(14.11) \qquad (1 - 2xt + t^2)^{-\frac{1}{2}} = \sum_{n=0}^{\infty} t^n P_n(x)$$

The function on the left side is called the *generating function* for the Legendre polynomials.

Let n be a non-negative integer. Then we have

$$\frac{d^n}{dx^n}(x^2 - 1)^n = \frac{d^n}{dx^n} \sum_{s=0}^{n} (-1)^s \frac{n!}{s! \, (n-s)!} x^{2n-2s}$$

$$= \sum_{0 \le s \le (n/2)} (-1)^s \frac{n!}{s!(n-s)!} \frac{(2n-2s)!}{(n-2s)!} x^{n-2s}$$

Thus we obtain *Rodrigue's formula*

$$(14.12) \qquad P_n(x) = \frac{1}{2^n n!} \frac{d^n}{dx^n}(x^2 - 1)^n$$

Orthogonality relations. By partial integration, we obtain

$$\int_{-1}^{1} \frac{d}{dx}\{(1 - x^2)y'\} z \, dx = [(1 - x^2)y' \, z]_{-1}^{1}$$

$$- \int_{-1}^{1} (1 - x^2) y' \, z' \, dx = \int_{-1}^{1} \frac{d}{dx}\{(1 - x^2) z'\} y \, dx$$

On the other hand, we have

$$\{(1 - x^2) P_n(x)'\}' = -n(n + 1) P_n(x)$$

Therefore,

$$-n(n + 1)\int_{-1}^{1} P_n(x) P_m(x) \, dx = -m(m + 1)\int_{-1}^{1} P_m(x) P_n(x) \, dx$$

from which we can derive the *orthogonality relations*

$$(14.13) \qquad \int_{-1}^{1} P_n(x) P_m(x) \, dx = 0 \qquad (n \ne m)$$

For $n = m$, we can prove

$$(14.14) \qquad \int_{-1}^{1} P_n(x)^2 \, dx = \frac{2}{2n + 1}$$

To show this, set

$$y_n = (x^2 - 1)^n$$

Then we see that

$$\int_{-1}^{1} y_n^{(n)} y_n^{(n)} \, dx = -\int_{-1}^{1} y_n^{(n-1)} y_n^{(n+1)} \, dx = \cdots$$

$$= (-1)^n \int_{-1}^{1} y_n y_n^{(2n)} \, dx$$

$$= (2n)! \int_{-1}^{1} (1 - x)^n (1 + x)^n \, dx$$

$$= \frac{n}{n + 1} (2n)! \int_{-1}^{1} (1 - x)^{n-1} (1 + x)^{n+1} \, dx$$

$$= \frac{n! \, (2n)!}{(n + 1)(n + 2) \cdots 2n} \int_{-1}^{1} (1 + x)^{2n} \, dx$$

$$= \frac{(n!)^2}{2n + 1} 2^{2n+1}$$

from which (14.14) can be derived immediately.

Completeness. The fact that every real-valued continuous function $f(x)$ defined on the interval $-1 \leqq x \leqq 1$ can be approximated by polynomials in x uniformly on this interval, is known as Weierstrass' theorem. Since $P_n(x)$ is a polynomial of degree n, $f(x)$ can be approximated by linear combinations

$$\sum_{j=1}^{n} C_j P_j(x)$$

uniformly on this interval. Having this property, the orthogonal system of the functions $\{P_n(x)\}$ is said to be *complete* on the interval $[-1, 1]$.

15. *Bessel differential equations*

The *Bessel differential equations* are derived from the Gauss differential equations as follows. By substituting kx for x in (13.1), the equation (13.1) becomes an equation of the form

$$x(1 - kx) y'' + (c - bx) y' - ay = 0$$

By letting $k \to 0$, this equation becomes

$$xy'' + (c - bx) y' - ay = 0$$

Then, by letting $b \to 0$, we obtain

$$xy'' + cy' - ay = 0$$

By replacing ax by x, this equation assumes the form

$$xy'' + cy' - y = 0$$

By the transformation $x = -t^2/4$ and by putting $c = \nu + 1$, the above equation becomes

$$ty_{tt} + (2\nu + 1)y_t + ty = 0$$

Further, by the transformation $y = t^{-\nu}z$, we obtain

$$t^2 z_{tt} + tz_t + (t^2 - \nu^2)z = 0$$

which is called Bessel's differential equation.

Thus Bessel's equation is defined by any one of the following three equations

(15.1) $$xy'' + (\nu + 1)y' - y = 0$$

(15.2) $$xy'' + (2\nu + 1)y' + xy = 0$$

(15.3) $$x^2 y'' + xy' + (x^2 - \nu^2)y = 0$$

We first consider the equation (15.3). The singular points for this equation are $x = 0$ and $x = \infty$. The point $x = 0$ is a regular singular point and $x = \infty$ an irregular singular point. The indicial equation at the point $x = 0$ is

(15.4) $$r^2 - \nu^2 = 0$$

and has roots ν and $-\nu$. First we shall consider the case when ν is not an integer. In this case, there exists a fundamental system of the solutions such that

$$y_1 = x^\nu \{a_0 + a_1 x + a_2 x^2 + \cdots\}, \qquad a_0 \neq 0$$
$$y_2 = x^{-\nu} \{b_0 + b_1 x + b_2 x^2 + \cdots\}, \qquad b_0 \neq 0$$

Let

$$y_1 = x^\nu z_1$$

Then by the above discussion, z_1 is a solution of the equation (15.2). Further, $z_1(-x)$ is a solution of the equation (15.2), which can be derived by easy calculation. Accordingly, if $a_1 = 0$, that is, $z_1'(0) = 0$, then, by the uniqueness of the solution, $z_1(x) \equiv z_1(-x)$. Therefore, we may assume, in this case $a_1 = 0$, that y_1 is of the form

$$y_1 = x^\nu \{c_0 + c_1 x^2 + c_2 x^4 + \cdots + c_\mu x^{2\mu} + \cdots\}$$

Substituting this in (15.3), we obtain, by comparison of coefficients, the following *recurrence relations*

(15.5) $$\{(\nu + 2\mu)^2 - \nu^2\} c_\mu = -c_{\mu-1}$$

Hence, setting

$$c_0 = \frac{1}{2^\nu \Gamma(\nu + 1)}$$

and by making use of the notation

$$\Gamma(\mu + 1) = \mu!$$

we see that the solution y_1 is given by

(15.6) $$J_\nu(x) = \sum_{\mu=0}^\infty \frac{(-1)^\mu}{\Gamma(\nu + \mu + 1)\,\Gamma(\mu + 1)} \left(\frac{x}{2}\right)^{\nu+2\mu}$$

which is called the *Bessel function* of the first kind of order ν, or *cylinder function*. For $-\nu$, we can obtain a solution

(15.7) $$J_{-\nu}(x) = \sum_{\mu=0}^\infty \frac{(-1)^\mu}{\Gamma(-\nu + \mu + 1)\,\Gamma(\mu + 1)} \left(\frac{x}{2}\right)^{-\nu+2\mu}$$

which is linearly independent of $J_\nu(x)$. Since the singular points for the equation (15.3) are only 0 and ∞, both series in (15.6) and (15.7) converge for $0 < |x| < \infty$.

Next, we shall consider the case when $\nu = n$ is a non-negative integer. In this case, $J_n(x)$ is well defined by (15.6) and satisfies the equation (15.3). To define $J_{-n}(x)$, it should be noted that $\Gamma(x)$ is a meromorphic function with simple poles at $x = 0, -1, -2, \cdots$ and no zeros. Because of this property, we can define $J_{-n}(x)$, letting $\nu \to n$ in (15.7), as follows:

(15.7)' $$J_{-n}(x) = \sum_{\mu=0}^\infty \frac{(-1)^{n+\mu}}{\Gamma(\mu + 1)\,\Gamma(n + \mu + 1)} \left(\frac{x}{2}\right)^{n+2\mu}$$

In this case, however, $J_{-n}(x)$ is linearly dependent of $J_n(x)$, since

(15.8) $$J_n(x) = (-1)^n J_{-n}(x)$$

Generating function. We consider the Laurent expansion of $\exp\left[\frac{1}{2}x(t - 1/t)\right]$ at the point $t = 0$

$$\exp\left[\tfrac{1}{2}x(t - 1/t)\right] = \sum_{n=-\infty}^\infty t^n f_n(x)$$

If we replace t by $-1/t$, then the right side becomes $\sum_{n=-\infty}^\infty (-t)^{-n} f_n(x)$ while the left side remains invariant. Hence, by the uniqueness of the expansion, we obtain

$$f_{-n}(x) = (-1)^n f_n(x)$$

On the other hand,

$$\exp\left(\frac{1}{2}x\left(t - \frac{1}{t}\right)\right) = \sum_{r=0}^{\infty} \frac{(\frac{1}{2}x)^r t^r}{r!} \sum_{m=0}^{\infty} \frac{(-\frac{1}{2}x)^m t^{-m}}{m!}$$

accordingly, the coefficient of t^n is

$$\sum_{m=0}^{\infty} \frac{(\frac{1}{2}x)^{n+m}}{(n+m)!} \frac{(-\frac{1}{2}x)^m}{m!}$$

Therefore, we obtain that, for $n \geqq 0$,

$$f_n(x) = J_n(x)$$

and accordingly

(15.9) $$\exp\left[\frac{1}{2}x\left(t - \frac{1}{t}\right)\right] = \sum_{n=-\infty}^{\infty} t^n J_n(x)$$

The function on the left side is called the generating function for the Bessel functions.

Neumann functions. The *Neumann functions*, or the *Bessel functions of the second kind*, are defined by

(15.10) $$Y_\nu(x) = \frac{J_\nu(x) \cos \nu\pi - J_{-\nu}(x)}{\sin \nu\pi}$$

If ν is not an integer, then obviously $Y_\nu(x)$ and $J_\nu(x)$ form a fundamental system. If ν is a non-negative integer, then, according to (15.8), (15.10) becomes an indeterminate form. To calculate the limit, differentiate both the denominator and the numerator with respect to ν. Then, setting $\nu = n$, we have

$$Y_n(x) = \lim_{\nu \to n} Y_\nu(x)$$
$$= \frac{1}{\pi}\left\{\frac{\partial J_\nu(x)}{\partial \nu}\right\}_{\nu=n} - \frac{(-1)^n}{\pi}\left\{\frac{\partial J_{-\nu}(x)}{\partial \nu}\right\}_{\nu=n}$$

Now set

(15.11) $$\psi(x) = \Gamma'(x)/\Gamma(x)$$

Then, by (15.6), we obtain that

$$\frac{\partial J_\nu(x)}{\partial \nu} = \sum_{m=0}^{\infty} \frac{(-1)^m(\frac{1}{2}x)^{\nu+2m}}{m!\,\Gamma(\nu+m+1)}\left\{\log\frac{x}{2} - \psi(\nu+m+1)\right\}$$

hence, as $\nu \to n$, $\partial J_\nu(x)/\partial\nu$ tends to

$$\sum_{m=0}^{\infty} \frac{(-1)^m (\frac{1}{2}x)^{n+2m}}{m!\,(n+m)!} \left\{ \log \frac{x}{2} - \psi(n+m+1) \right\}$$

We now write

$$J_{-\nu}(x) = \sum_{m=0}^{n-1} \frac{(-1)^m (\frac{1}{2}x)^{-\nu+2m}}{m!\,\Gamma(-\nu+m+1)} + \sum_{m=n}^{\infty} \frac{(-1)^m (\frac{1}{2}x)^{-\nu+2m}}{m!\,\Gamma(-\nu+m+1)}$$

By making use of the formula

$$(15.12) \qquad \Gamma(x)\,\Gamma(1-x) = \frac{\pi}{\sin \pi x}$$

we obtain

$$\frac{(\frac{1}{2}x)^{-\nu+2m}}{\Gamma(-\nu+m+1)} = \frac{(\frac{1}{2}x)^{-\nu+2m}\,\Gamma(\nu-m)\sin(\nu-m)\pi}{\pi}$$

Hence, differentiating it, we see that, for $0 \leqq m < n$,

$$\left[\frac{\partial}{\partial \nu} \left\{ \frac{(\frac{1}{2}x)^{-\nu+2m}\,\Gamma(\nu-m)\sin(\nu-m)\pi}{\pi} \right\} \right]_{\nu=n}$$
$$= [(\tfrac{1}{2}x)^{-\nu+2m}\,\Gamma(\nu-m)\{\pi^{-1}\psi(\nu-m)\sin(\nu-m)\pi$$
$$+ \cos(\nu-m)\pi - \pi^{-1}\log(\tfrac{1}{2}x)\sin(\nu-m)\pi\}]_{\nu=n}$$
$$= (\tfrac{1}{2}x)^{-n+2m}\,\Gamma(n-m)\cos(n-m)\pi$$

Therefore, as $\nu \to n$, $\partial J_{-\nu}(x)/\partial \nu$ tends to

$$\sum_{m=0}^{n-1} \frac{(-1)^n\,\Gamma(n-m)(\frac{1}{2}x)^{-n+2m}}{m!}$$
$$+ \sum_{m=n}^{\infty} \frac{(-1)^m (\frac{1}{2}x)^{-n+2m}}{m!\,(-n+m)!}\{-\log(\tfrac{1}{2}x) + \psi(-n+m+1)\}$$
$$= (-1)^n \sum_{m=0}^{n-1} \frac{(n-m-1)!}{m!}(\tfrac{1}{2}x)^{-n+2m}$$
$$+ (-1)^{n-1} \sum_{m=0}^{\infty} \frac{(-1)^m (\frac{1}{2}x)^{n+2m}}{m!\,(n+m)!}\{\log(\tfrac{1}{2}x) - \psi(m+1)\}$$

To sum up, the Neumann function $Y_n(x)$ with n being a non-negative integer is given by

$$Y_n(x) = \frac{2}{\pi} J_n(x) \log\left(\frac{x}{2}\right)$$

$$(15.13) \qquad - \frac{1}{\pi} \sum_{m=0}^{\infty} (-1)^m \frac{(\psi(m+1) + \psi(n+m+1))}{m!(n+m)!} \left(\frac{x}{2}\right)^{n+2m}$$

$$- \frac{1}{\pi} \sum_{m=0}^{n-1} \frac{(n-1-m)!}{m!} \left(\frac{x}{2}\right)^{-n+2m}$$

where the last term does not appear when $n = 0$. (15.13) is known as *Hankel's formula*.

Since $Y_\nu(x)$ is a solution of the equation (15.3) containing a parameter ν, and, as $\nu \to n$, $Y_\nu(x)$ converges uniformly to $Y_n(x)$ in any bounded closed domain in the complex x-plane except for the origin, $Y_n(x)$ is a solution of (15.3) for $\nu = n$. It is easily seen that $Y_n(x)$ and $J_n(x)$ form a fundamental system.

Lommel's formula. As is proved above, $Y_\nu(x)$ and $J_\nu(x)$ form a fundamental system for every ν. We shall prove here that the Wronskian $W(J_\nu(x), Y_\nu(x))$ satisfies

(15.14) $\qquad W(J_\nu(x), Y_\nu(x)) = J_\nu(x) Y_\nu'(x) - J_\nu'(x) Y_\nu(x) = 2/(\pi x)$

(15.14) is known as *Lommel's formula*.

Proof of (15.14). Multiply

$$J_\nu'' + \frac{1}{x} J_\nu' + \left(1 - \frac{\nu^2}{x^2}\right) J_\nu = 0$$

and

$$J_{-\nu}'' + \frac{1}{x} J_{-\nu}' + \left(1 - \frac{\nu^2}{x^2}\right) J_{-\nu} = 0$$

by $J_{-\nu}$ and J_ν respectively. Then, by subtraction, we obtain

$$\frac{d}{dx}\{J_\nu' J_{-\nu} - J_\nu J_{-\nu}'\} + \frac{1}{x}\{J_\nu' J_{-\nu} - J_\nu J_{-\nu}'\} = 0$$

and from this,

$$J_\nu'(x) J_{-\nu}(x) - J_\nu(x) J_{-\nu}'(x) = C/x$$

The integration constant C is determined as follows. From (15.6), it follows that, as $x \to 0$,

$$J_\nu(x) = \frac{(\tfrac{1}{2}x)^\nu}{\Gamma(\nu + 1)}(1 + O(x^2))$$

$$J_\nu'(x) = \frac{(\tfrac{1}{2}x)^{\nu-1}}{2\Gamma(\nu)}(1 + O(x^2))$$

Hence, by making use of (15.12), we see that

$$W(J_\nu(x), J_{-\nu}(x)) = \frac{1}{x}\left\{\frac{1}{\Gamma(\nu + 1)\Gamma(-\nu)} - \frac{1}{\Gamma(\nu)\Gamma(-\nu + 1)}\right\} + O(x)$$

$$= -\frac{2\sin \nu\pi}{\pi x} + O(x)$$

from which we can derive

$$C = - \frac{2 \sin \nu\pi}{\pi}$$

Therefore we obtain

(15.15) $$W(J_\nu(x), J_{-\nu}(x)) = - \frac{2 \sin \nu\pi}{\pi x}$$

Introducing (15.10) we obtain further

$$W(J_\nu(x), Y_\nu(x)) = \frac{\cos \nu\pi}{\sin \nu\pi} W(J_\nu(x), J_\nu(x)) - \frac{1}{\sin \nu\pi} W(J_\nu(x), J_{-\nu}(x))$$

$$= \frac{2}{\pi x}$$

when ν is not an integer. Since $W(J_\nu(x), Y_\nu(x))$ is continuous with respect to ν, the above relation holds for $\nu = n$, n being an integer. Consequently, (15.14) holds for every ν, q.e.d.

Hankel functions. The *Hankel functions*, or, the *Bessel functions of the third kind* are defined by

(15.16)
$$H_\nu^{(1)}(x) = J_\nu(x) + i Y_\nu(x)$$
$$= \frac{J_{-\nu}(x) - e^{-\nu\pi i} J_\nu(x)}{i \sin \nu\pi}$$
$$H_\nu^{(2)}(x) = J_\nu(x) - i Y_\nu(x)$$
$$= \frac{J_{-\nu}(x) - e^{\nu\pi i} J_\nu(x)}{-i \sin \nu\pi} \qquad (i = \sqrt{-1})$$

We shall discuss here the case when $\nu = \frac{1}{2}$, to illustrate the relation between the Bessel, the Neumann, and the Hankel functions. By making use of

$$\Gamma(x + 1) = x\Gamma(x) , \qquad \Gamma(\tfrac{1}{2}) = \sqrt{\pi}$$

we see that

(15.17) $$J_{-\frac{1}{2}}(x) = \frac{(\tfrac{1}{2}x)^{-\frac{1}{2}}}{\Gamma(\tfrac{1}{2})} \sum_{m=0}^\infty \frac{(-1)^m (\tfrac{1}{2}x)^{2m}}{m! (\tfrac{1}{2})(\tfrac{3}{2}) \cdots (m - \tfrac{1}{2})} = (2/\pi x)^{\frac{1}{2}} \cos x$$

(15.18) $$J_{\frac{1}{2}}(x) = \frac{(\tfrac{1}{2}x)^{\frac{1}{2}}}{\Gamma(\tfrac{1}{2})} \sum_{m=0}^\infty \frac{(-1)^m (\tfrac{1}{2}x)^{2m}}{m! (m + \tfrac{1}{2})(m - \tfrac{1}{2}) \cdots (\tfrac{1}{2})} = (2/\pi x)^{\frac{1}{2}} \sin x$$

Hence we obtain from (15.10)

(15.19) $$Y_{\frac{1}{2}}(x) = -J_{-\frac{1}{2}}(x), \ Y_{-\frac{1}{2}}(x) = J_{\frac{1}{2}}(x)$$

so that

(15.20)
$$H_{\frac{1}{2}}^{(1)}(x) = -i(2/\pi x)^{\frac{1}{2}} e^{ix}$$
$$H_{\frac{1}{2}}^{(2)}(x) = i(2/\pi x)^{\frac{1}{2}} e^{-ix}$$

In general, as is suggested by the above results, it is known that the relation between the Bessel, the Neumann, and the Hankel functions is similar as that between sine, cosine, and exponential functions. However, the analogy has nothing to do with the following, so is omitted. We shall here prove the following fact which will be needed in Part 52.

(15.21)   There exist a positive number $\rho(\infty)$ and a real number $\theta_1(\infty)$ such that

$$J_n(x) \approx \frac{\rho(\infty)}{\sqrt{x}} \sin (x + \theta_1(\infty))$$

as $x \to \infty$.

Proof of (15.21).   Set

(15.22) $$y(x) = x^{\frac{1}{2}} J_n(x)$$

Since

$$x^2 J_n'' + x J_n' + (x^2 - n^2) J_n = 0$$

$y(x)$ satisfies

(15.23) $$y'' + \left\{ 1 - \frac{n^2 - \frac{1}{4}}{x^2} \right\} y = 0$$

According to (15.22), $y(x)$ is real for any positive number x and not identically zero. Further, by (15.23), either $y(x)$ or $y'(x)$ does not vanish whenever x is a positive number. Hence we may write

(15.24)
$$y(x) = \rho(x) \sin \theta(x)$$
$$y'(x) = \rho(x) \cos \theta(x)$$
$$\rho(x) = \{ y(x)^2 + y'(x)^2 \}^{\frac{1}{2}} > 0$$

Then, by (15.23), $\rho(x)$ and $\theta(x)$ must satisfy

(15.25)
$$\rho' = \frac{n^2 - \frac{1}{4}}{x^2} \rho \sin \theta \cos \theta$$
$$\theta' = 1 - \frac{n^2 - \frac{1}{4}}{x^2} \sin^2 \theta$$

From the first equation, we obtain

$$0 < \rho(x) = \rho(x_0) \exp\left(\int_{x_0}^{x} \frac{n^2 - \frac{1}{4}}{t^2} \sin\theta(t)\cos\theta(t)\,dt\right)$$

Since

$$|\sin\theta(t)\cos\theta(t)| \leqq 1$$

we see that

$$\lim_{x\to\infty} \rho(x) = \rho(\infty) < \infty$$

and $\rho(\infty) > 0$.

Substituting $x + \theta_1(x)$ for $\theta(x)$ in the second equation in (12.25), we obtain

$$\theta_1' = \frac{\frac{1}{4} - n^2}{x^2} \sin^2(\theta_1 + x)$$

Hence we obtain

$$\theta_1(x) - \theta_1(x_0) = \int_{x_0}^{x} \frac{\frac{1}{4} - n^2}{t^2} \sin^2(t + \theta_1(t))\,dt$$

Therefore, we see that

$$\lim_{x\to\infty} \theta_1(x) = \theta_1(\infty) < \infty$$

Thus, the *asymptotic formula* (15.21) of the Bessel functions is proved.

Concerning further details of the theory of the Bessel functions, the reader is referred to G. N. Watson, *A Treatise on the Theory of Bessel Functions*, Cambridge, 1922.

CHAPTER 2

THE BOUNDARY VALUE PROBLEM FOR LINEAR DIFFERENTIAL EQUATIONS OF SECOND ORDER

§ 1. Boundary value problem

16. *Boundary value problem of Sturm-Liouville type*

We consider a differential equation of the second order

$$(16.1) \qquad y'' + p_1(x)y' + p_2(x)y = 0$$

where $p_1(x)$ and $p_2(x)$ are real-valued continuous functions on a finite closed interval $a \leqq x \leqq b$. Multiply both sides by

$$(16.2) \qquad \exp\left(\int_a^x p_1(t)dt\right) = p(x)$$

Then the equation (16.1) assumes the form

$$(16.3) \qquad \frac{d}{dx}\left(p(x)\frac{dy}{dx}\right) = q(x)y$$

where

$$q(x) = -p_2(x)p(x)$$

The coefficients $p(x)$ and $q(x)$ satisfy the following conditions:

(16.4) $p(x)$ and $q(x)$ are real-valued continuous functions on the interval $a \leqq x \leqq b$, and $p(x) > 0$ there.

Conversely, if the coefficients $p(x)$ and $q(x)$ satisfy the condition (16.4) and if $p(x)$ is continuously differentiable, then the equation (16.3) becomes an equation of the form (16.1). In the following, we shall be concerned with the equation (16.3) under the assumption (16.4). Such an extended equation (16.3) is equivalent to a system of the equations

$$(16.3') \qquad dy/dx = z/p(x)\,, \quad dz/dx = q(x)y$$

to which the existence and uniqueness theorem in Part 7 can be immediately applied. If a pair of functions $y(x)$ and $z(x)$ is a solution of the equations (16.3') and if $y(x) \not\equiv 0$, then, according to the uniqueness theorem, the functions $y(x)$ and $z(x)$ do not vanish simultaneously at any point x in the interval $[a, b]$. Therefore, as far as we are concerned with the solution $y(x) \not\equiv 0$ of (16.3'), we may set

(16.5)
$$\begin{cases} y(x) = \rho(x) \sin \theta(x) \\ z(x) = \rho(x) \cos \theta(x) \\ \rho(x) = (y(x)^2 + z(x)^2)^{1/2} > 0 \end{cases}$$

and transform the system of the equations (16.3′) to a system of the equations, in unknowns ρ and θ,

(16.6)
$$\rho' = \left(\frac{1}{p(x)} + q(x) \right) \rho \sin \theta \cos \theta$$

$$\theta' = \frac{1}{p(x)} \cos^2\theta - q(x) \sin^2\theta , \quad \rho(x) > 0$$

The second equation does not contain the unknown ρ, hence we can find a solution $\theta(x)$. Then, substituting this solution in the first equation, we can obtain the general solution $\rho(x)$

$$\rho(x) = \rho(a) \exp\left(\int_a^x \left\{ \frac{1}{p(t)} + q(t) \right\} \sin \theta(t) \cos \theta(t) dt \right)$$

Since $\rho(x) > 0$ or < 0 for every point $x \in [a, b]$ according as $\rho(a) > 0$ or < 0, we can find a positive solution $\rho(x)$ from which, combined with the solution $\theta(x)$, we can derive a solution $y(x) = \rho(x) \sin \theta(x)$, not identically zero, of the original equation (16.3).

For every integer n, $\theta(x) + 2n\pi$ is also a solution of the second equation in (16.6). Furthermore the right side of the first equation in (16.6) is a periodic function of θ with period 2π. Hence we obtain the following theorem.

THEOREM 16.1   By means of a solution $\theta(x)$ of the second equation of (16.6), a positive solution $\rho(x)$ of the first equation is uniquely determined up to the multiplicative factor, so that the corresponding solution $y(x)$ of the original equation is also uniquely determined up to the multiplicative factor. If $y_1(x)$ and $y_2(x)$ are the solutions corresponding to the solutions $\theta_1(x)$ and $\theta_2(x) = \theta_1(x) + 2n\pi$ respectively, then $y_1(x)$ and $y_2(x)$ are linearly dependent.

Conversely, if two solutions $y_1(x) = \rho_1(x) \sin \theta_1(x)$ and $y_2(x) = \rho_2(x) \sin \theta_2(x)$ are linearly dependent, then, for some integer n, $\theta_1(x) = \theta_2(x) + 2n\pi$.

Proof. Let $y_1(x) = Cy_2(x)$. Then $y_1'(x) = Cy_2'(x)$. Hence

$$z_1(x) = p(x)y_1'(x) = Cp(x)y_2'(x) = Cz_2(x)$$

so that $\rho_1(x) = C\rho_2(x)$. Therefore we obtain $\theta_1(x) = \theta_2(x) + 2n\pi$, since $y_1(x) = Cy_2(x)$, and $z_1(x) = Cz_2(x)$,   q.e.d.

Accordingly, the linear dependence of two solutions $y_1(x) \not\equiv 0$ and $y_2(x) \not\equiv 0$ of the original equation (16.3) corresponds to the congruence relation

(16.7) $$\theta_1(x) \equiv \theta_2(x) \qquad (\mathrm{mod}\ 2\pi)$$

between the corresponding solutions $\theta_1(x)$ and $\theta_2(x)$ of the equations (16.6). Further, an initial condition for $\theta(x)$,

(16.8) $$\theta(a) = \alpha$$

corresponds to a linear homogeneous relation

(16.9) $$p(a)y'(a) \sin \alpha - y(a) \cos \alpha = 0$$

between $y(a)$ and $y'(a)$, as is easily seen from (16.3') and (16.5). We also see that the initial conditions

(16.10) $$y(a) = y_0\ , \quad y'(a) = y'_0$$

which were dealt with in Chapter 1, correspond to initial conditions

(16.10') $$\theta(a) = \alpha\ , \quad \rho(a) = \rho_0$$

for the corresponding solutions $\theta(x)$ and $\rho(x)$. In this section we shall be concerned with the problem of finding the solution $y(x)$ corresponding to the solution $\theta(x)$ satisfying the *boundary conditions*

(16.11) $$\theta(a) = \alpha\ , \quad \theta(b) = \beta$$

at both end points a and b of the interval $[a, b]$. Conditions (16.11) correspond to conditions

(16.11')
$$p(a)y'(a) \sin \alpha - y(a) \cos \alpha = 0$$
$$p(b)y'(b) \sin \beta - y(b) \cos \beta = 0$$

for $y(x)$. Hence, in general, the solution $y(x)$ of the equation (16.3) satisfying the conditions (16.11') is not uniquely determined. In fact, if $y_1(x)$ and $y_2(x)$ satisfy (16.3) and (16.11'), then, for any constants $C_1,\ C_2$,

$$C_1 y_1(x) + C_2 y_2(x)$$

also satisfies (16.3) and (16.11'). Thus the *boundary value problem*, which consists in finding a solution of the equation (16.3) satisfying the linear homogeneous relations (16.11') between y and y', is essentially different from the initial value problem (16.3), (16.10).

We now consider an equation, containing a complex parameter λ,

(16.12) $$(p(x)y')' - q(x)y + \lambda r(x)y = 0$$

where $r(x)$ is a function continuous and positive on the interval $[a, b]$. The *boundary value problem of Sturm-Liouville type* consists in finding a *non-trivial solution* $y(x)$, that is, $y(x) \not\equiv 0$, of the equation (16.12) satisfying the boundary conditions (16.11′). In general, the solution does not always exist for all values of parameter λ. For example, consider the boundary value problem

$$y'' + \lambda y = 0, \, y(0) = y(1) = 0$$

The solution exists only for $\lambda = (n\pi)^2$, $n = 1, 2, \cdots$, and is $\sin(n\pi x)$, $n = 1, 2, \cdots$. The *eigenvalue problem* is to determine values of complex parameter λ for which the solution of the prescribed boundary value problem exists. The value of such parameter is called an *eigenvalue*, and the corresponding solution is called an *eigenfunction* belonging to the eigenvalue. In the following, we shall show that the eigenvalue problem can be reduced to an integral equation with symmetric kernel.

17. Green's function. Reduction to integral equations

We denote by $L_x(y)$ a differential operator

$$(17.1) \qquad L_x(y) = \frac{d}{dx}\left(p(x)\frac{dy}{dx}\right) - q(x)y$$

which is defined for every function $y(x)$ such that dy/dx and $d/dx \cdot (p(x)(dy/dx))$ are defined and continuous on the interval $[a, b]$. Integrating both sides of *Lagrange's identity*

$$(17.2) \qquad yL_x(z) - zL_x(y)$$
$$= \frac{d}{dx}p(x)[y(x)z'(x) - y'(x)z(x)]$$

we obtain

$$(17.3) \quad [p(x)(y(x)z'(x) - y'(x)z(x))]_{a'}^{b'}$$
$$= \int_{a'}^{b'} \{y(x)L_x(z) - z(x)L_x(y)\}dx \, , \quad a \leqq a' < b' \leqq b$$

which is known as *Green's formula*.[1]

---

[1] In the case when $L_x(y)$ and $L_x(z)$ are not defined on the whole interval $[a, b]$ but only on an open interval (a', b'), $a \leqq a' < b' \leqq b$, the formula (17.3) also holds true, if all of $y(x)$, $y'(x)$, $z(x)$, and $z'(x)$ have finite limits as $x \downarrow a'$ and $x \uparrow b'$ and if we denote

$$[\quad]_{a'}^{b'} = \lim_{a_1' \downarrow a', b_1' \uparrow b'} [\quad]_{a_1'}^{b_1'} = [\quad]_{a'+0}^{b'-0}$$

$$\int_{a'}^{b'} = \lim_{a_1' \downarrow a', b_1' \uparrow b'} \int_{a_1'}^{b_1'} = \int_{a'+0}^{b'-0}$$

In particular, if two functions $y(x)$ and $z(x)$ both satisfy the boundary conditions (16.11'), then, setting $a = a'$ and $b = b'$, we see that the left side of (17.3) becomes zero. Hence we have the following:

(17.4) If two functions $y(x)$ and $z(x)$ both satisfy the boundary conditions (16.11'), then

$$\int_a^b \{y(x)L_x(z) - z(x)L_x(y)\}dx = 0$$

Green's function. Suppose that two functions $y_1(x) \not\equiv 0$ and $y_2(x) \not\equiv 0$ satisfy

(17.5)
$$L_x(y_1) = 0$$
$$p(a)y_1'(a)\sin\alpha - y_1(a)\cos\alpha = 0$$

and

(17.6)
$$L_x(y_2) = 0$$
$$p(b)y_2'(b)\sin\beta - y_2(b)\cos\beta = 0$$

respectively, and suppose that these two functions $y_1(x)$ and $y_2(x)$ are linearly independent. Write

$$C = p(\xi)[y_1(\xi)y_2'(\xi) - y_1'(\xi)y_2(\xi)]$$

Differentiating C with respect to ξ and making use of (17.2), we see, by virtue of (17.5) and (17.6), that C must be a constant. Moreover, the linear independence of $y_1(x)$ and $y_2(x)$ implies that C is not zero. Now we define a function $G(x, \xi)$ of two variables x and ξ by

(17.7)
$$G(x, \xi) = -\frac{1}{C}y_1(\xi)y_2(x) \quad (x \geq \xi)$$
$$= -\frac{1}{C}y_1(x)y_2(\xi) \quad (x < \xi)$$

$$C = p(\xi)[y_1(\xi)y_2'(\xi) - y_1'(\xi)y_2(\xi)] \equiv \text{constant}$$

The function $G(x, \xi)$ is called *Green's function* for the equation $L_x(y) = 0$ subject to the boundary conditions (16.11').

Obviously, Green's function $G(x, \xi)$ has the following properties:

(17.8) $G(x, \xi)$ is continuous at any point (x, ξ) in the domain $a \leq x, \xi \leq b$. As a function of x, $G(x, \xi)$ satisfies the given boundary conditions for every ξ,

(17.9) If $x \neq \xi$, $G(x, \xi)$ satisfies the equation $L_x(G) = 0$ as a function of x. Both $G_x(x, \xi)$ and $\{p(x)G_x(x, \xi)\}_x$ are bounded in the region $x \neq \xi$, $a \leq x, \xi \leq b$,[1]

----

[1] $G_x(x, \xi) = \partial G(x, \xi)/\partial x$.

(17.10)   If $a < x_0 < b$, then as $x \to x_0, \xi \to x_0$ keeping the relation $x > \xi$ and as $x \to x_0, \xi \to x_0$ keeping the relation
$x < \xi$, $G_x(x, \xi)$ tends to finite values $G_x(x_0 + 0, x_0)$ and
$G_x(x_0 - 0, x_0)$, respectively, and
$G_x(x_0 + 0, x_0) - G_x(x_0 - 0, x_0) = (-1/p(x_0))$, [1]

(17.11) $G(x, \xi) \equiv G(\xi, x)$

Reduction to integral equations.   By making use of Green's function and its properties (17.8)–(17.11), we shall prove the following
two theorems.   We start with

THEOREM 17.1.   Let $\varphi(x)$ be a continuous function defined on the
interval $[a, b]$.   If $y(x)$ is a solution of the equation

(17.12) $L_x(y) = -\varphi(x)$

satisfying the boundary conditions (16.11′), then $y(x)$ can be written
in the form

(17.13) $y(x) = \int_a^b G(x, \xi)\varphi(\xi)d\xi$

Proof.   Applying Green's formula to the functions $y(x)$ and $z(x) =$
$G(x, \xi)$, we obtain

$$\int_a^{\xi-0} \{y(x)L_x(G(x, \xi)) - G(x, \xi)L_x(y)\}dx$$

$$+ \int_{\xi+0}^b \{y(x)L_x(G(x, \xi)) - G(x, \xi)L_x(y)\}dx$$

$$= \left[p(x)\left(y(x)\frac{\partial G(x, \xi)}{\partial x} - y'(x)G(x, \xi)\right)\right]_a^{\xi-0}$$

$$+ \left[p(x)\left(y(x)\frac{\partial G(x, \xi)}{\partial x} - y'(x)G(x, \xi)\right)\right]_{\xi+0}^b$$

for every $\xi, a < \xi < b$.   The equation (17.12), together with (17.9),
implies that the left side is equal to

$$\int_a^b G(x, \xi)\varphi(x)dx$$

On the other hand, (17.8), together with (17.10), implies that the
right side is equal to $y(\xi)$.   Therefore, we obtain

[1] We note that we may equally well write $G_x(x_0, x_0 - 0)$ and $G_x(x_0, x_0 + 0)$
for $G_x(x_0 + 0, x_0)$ and $G_x(x_0 - 0, x_0)$, respectively.

$$y(\xi) = \int_a^b G(x, \xi)\varphi(x)dx \quad (a < \xi < b)$$

Since the two functions on both sides are continuous for $a \leqq \xi \leqq b$, the above identity also holds for $\xi = a$ and $\xi = b$. This, together with (17.11), proves the theorem, q.e.d.

Next we shall prove the following

THEOREM 17.2. The function $y(x)$ given by (17.13) satisfies both the equation (17.12) and the boundary conditions (16.11′).

Proof. By (17.7), we have

$$y(x) = -\frac{1}{C}y_2(x)\int_a^x y_1(\xi)\varphi(\xi)d\xi$$
$$-\frac{1}{C}y_1(x)\int_x^b y_2(\xi)\varphi(\xi)d\xi$$

Hence we obtain

$$y'(x) = -\frac{1}{C}y_2'(x)\int_a^x y_1(\xi)\varphi(\xi)d\xi - \frac{1}{C}y_2(x)y_1(x)\varphi(x)$$
$$-\frac{1}{C}y_1'(x)\int_x^b y_2(\xi)\varphi(\xi)d\xi + \frac{1}{C}y_1(x)y_2(x)\varphi(x)$$
$$= -\frac{1}{C}y_2'(x)\int_a^x y_1(\xi)\varphi(\xi)d\xi - \frac{1}{C}y_1'(x)\int_x^b y_2(\xi)\varphi(\xi)d\xi$$
$$= \int_a^{x-0} G_x(x, \xi)\varphi(\xi)d\xi + \int_{x+0}^b G_x(x, \xi)\varphi(\xi)d\xi$$
$$= \int_a^b G_x(x, \xi)\varphi(\xi)d\xi \; [1]$$

Similarly, we obtain, from the above,

$$(p(x)y'(x))'$$
$$= \int_a^b \{p(x)G_x(x, \xi)\}_x\varphi(\xi)d\xi + p(x)[G_x(x, \xi)\varphi(\xi)]_{\xi=x-0}^{\xi=x+0}$$

On account of (17.10), the second term on the right side equals $-\varphi(x)$. Hence we obtain, by (17.9),

$$L_x(y) = \int_a^b L_x(G(x, \xi))\varphi(\xi)d\xi - \varphi(x) = -\varphi(x)$$

q. e. d.

[1] $G_x(x, \xi)$ is discontinuous for $x = \xi$, but it does not matter for this integration because of (17.9).

Accordingly, in the case when there exist two linearly independent solutions y_1 and y_2, given by (17.5) and (17.6) respectively, we can reduce, by means of Green's function (17.7), our boundary value problem, (16.12) and (16.11′), to the integral equation

$$(17.14) \qquad y(x) = \lambda \int_a^b G(x, \xi) r(\xi) y(\xi) d\xi$$

When (17.14) has a non-trivial solution $y(x) \not\equiv 0$ for some λ, then λ is an eigenvalue for the boundary value problem, (16.12) and (16.11′), and the corresponding solution $y(x)$ is an eigenfunction belonging to this eigenvalue.

Examples of Green's functions. We have, on the basis of (17.7), the following examples of Green's functions.

1. $L_x = \dfrac{d^2}{dx^2}$, $y(0) = y(1) = 0$:

$$G(x, \xi) = \begin{cases} (1 - \xi)x , & (x \leq \xi) \\ (1 - x)\xi , & (x > \xi) \end{cases}$$

2. $L_x = \dfrac{d^2}{dx^2}$, $y(0) = y'(1) = 0$:

$$G(x, \xi) = \begin{cases} x , & (x \leq \xi) \\ \xi , & (x > \xi) \end{cases}$$

3. $L_x = \dfrac{d^2}{dx^2}$, $y(0) = y'(0), y(1) = -y'(1)$:

$$G(x, \xi) = \begin{cases} -\dfrac{1}{3}(x + 1)(\xi - 2) & (x \leq \xi) \\ -\dfrac{1}{3}(\xi + 1)(x - 2) , & (x > \xi) \end{cases}$$

18. *Periodic solutions. Generalized Green's function*

A system of important boundary conditions not included in the boundary conditions (16.11′) is as follows:

$$(18.1) \qquad y(a) = y(b) , \quad y'(a) = y'(b) \cdot$$

If the coefficients $p(x)$, $q(x)$ and $r(x)$ are *periodic functions* with period $b - a$, that is,

$$p(x + b - a) \equiv p(x), \quad q(x + b - a) \equiv q(x), \quad r(x + b - a) \equiv r(x)$$

then, the conditions (18.1) are just the conditions that the solution $y(x)$ of the equation (16.12) is periodic with the same period $b - a$, that is,

$$y(x + b - a) \equiv y(x)$$

For, in such a case, $y(x)$ and $y_{a,b}(x) = y(x + b - a)$ both satisfy the equation (16.12) together with the same initial conditions

$$y(a) = y_{a,b}(a), \quad y'(a) = y'_{a,b}(a)$$

Hence, by the uniqueness of the solutions, we must have

$$y(x) \equiv y_{a,b}(x)$$

In the following, we shall be concerned with more general conditions, which include the conditions (18.1), of the form

$$(18.2) \qquad y(a) = \gamma y(b), \ p(a)y'(a) = \frac{p(b)}{\gamma}y'(b)$$

or

$$(18.3) \qquad y(a) = \gamma p(b) y'(b), \ p(a)y'(a) = -\frac{1}{\gamma}y(b)$$

where γ is a non-zero constant.

As in the case of (16.11'), it is easily seen that if $y(x)$ and $z(x)$ both satisfy either (18.2) or (18.3), then the relation

$$(18.4) \qquad [p(x)(y(x)z'(x) - y'(x)z(x))]_a^b = 0$$

holds.

For the present case, we cannot use (17.7) to construct a function $G(x, \xi)$ satisfying (17.8)–(17.11) pertaining to the boundary conditions (18.2) or (18.3). However, if we can find, in some way, such a function $G(x, \xi)$—we shall also call it the Green's function for $L_x(y)$ under the boundary conditions (18.2) or (18.3)—then we can also reduce our boundary value problem to an integral equation. We have to prove the following: Every function satisfying (16.12) and (18.2) (or (18.3)) can be written in the form (17.14). Conversely, every function written as (17.14) satisfies (16.12) and (18.2) (or (18.3)). The first half can be proved as in Part 17. To prove the converse, differentiate both sides of (17.13), keeping in mind that $G_x(x, \xi)$ is not continuous at the point $x = \xi$. Then we obtain

$$\frac{dy}{dx} = \int_a^{x-0} \frac{\partial G(x, \xi)}{\partial x} \varphi(\xi)d\xi + \int_{x+0}^b \frac{\partial G(x, \xi)}{\partial x} \varphi(\xi)d\xi$$

Hence, for $\delta > 0$, we obtain

$$\frac{p(x + \delta)y'(x + \delta) - p(x)y'(x)}{\delta}$$

$$= \int_a^x \frac{1}{\delta}\{p(x + \delta)G_x(x + \delta, \xi) - p(x)G_x(x, \xi)\}\,\varphi(\xi)\,d\xi$$

$$+ \frac{1}{\delta}\int_x^{x+\delta} p(x + \delta)G_x(x + \delta, \xi)\,\varphi(\xi)\,d\xi$$

$$- \frac{1}{\delta}\int_x^{x+\delta} p(x)G_x(x, \xi)\varphi(\xi)d\xi$$

$$+ \int_{x+\delta}^b \frac{1}{\delta}\{p(x + \delta)G_x(x + \delta, \xi) - p(x)G_x(x, \xi)\}\varphi(\xi)d\xi$$

When $\delta \downarrow 0$, this converges to

$$\int_a^x \{p(x)G_x(x, \xi)\}_x\varphi(\xi)d\xi + p(x)\{G_x(x + 0, x) - G_x(x, x + 0)\}\varphi(x)$$

$$+ \int_x^b \{p(x)G_x(x, \xi)\}_x\varphi(\xi)d\xi$$

Similarly we can prove that, when $\delta \uparrow 0$,

$$\frac{p(x + \delta)y'(x + \delta) - p(x)y'(x)}{\delta}$$

also converges to the same limit. Hence the first derivative of $p(x)y'(x)$ exists and

$$(p(x)y'(x))' = \int_a^b \{p(x)G_x(x, \xi)\}_x\varphi(\xi)d\xi - \varphi(x)$$

Thus, by virtue of $L_x(G(x, \xi)) = 0$, $y(x)$ is a solution of the equation $L_x(y) = -\varphi(x)$. Since $G(x, \xi)$ satisfies the prescribed boundary conditions for every $\xi, a < \xi < b$, $y(x)$ also satisfies the boundary conditions, q.e.d.

Construction of Green's function. Suppose that a Green's function $G(x, \xi)$ exists. Then, since $L_x(G(x, \xi)) = 0$ for $x \neq \xi$, $G(x, \xi)$ must be represented, by means of a fundamental system $Y_1(x), Y_2(x)$ of the solutions of the equation $L_x(y) = 0$, as follows:

$$(18.5) \qquad G(x, \xi) = \begin{cases} C_1 Y_1(x) + C_2 Y_2(x), & (a \leqq x < \xi) \\ C_3 Y_1(x) + C_4 Y_2(x), & (\xi < x \leqq b) \end{cases}$$

where every C_i is a function $C_i(\xi)$ of ξ. We shall determine the relations between C_i so that $G(x, \xi)$ satisfies the required properties for Green's function pertaining to the boundary conditions (18.2). Since $G(x, \xi)$ is continuous at $x = \xi$, we obtain

$$(18.6) \qquad C_1 Y_1(\xi) + C_2 Y_2(\xi) = C_3 Y_1(\xi) + C_4 Y_2(\xi)$$

By (17.10), we obtain

$$(18.7) \qquad C_1 Y_1'(\xi) + C_2 Y_2'(\xi) - C_3 Y_1'(\xi) - C_4 Y_2'(\xi) = \frac{1}{p(\xi)}$$

Finally, from the boundary conditions (18.2), we obtain

$$(18.8) \qquad \begin{cases} C_1 Y_1(a) + C_2 Y_2(a) = \gamma(C_3 Y_1(b) + C_4 Y_2(b)) \\ \gamma p(a)\{C_1 Y_1'(a) + C_2 Y_2'(a)\} = p(b)\{C_3 Y_1'(b) + C_4 Y_2'(b)\} \end{cases}$$

Only the last relation (18.8) must be changed according as the corresponding boundary conditions, if we are concerned with Green's function under the boundary conditions (18.3).

If there exists one and only one set of functions C_1, C_2, C_3, and C_4 which satisfy the conditions (18.6)–(18.8), then the function, given by (18.5), is no other than the required Green's function, provided that (18.5) satisfies the symmetry condition $G(x, \xi) = G(\xi, x)$. However, if these functions C_1, C_2, C_3, and C_4 are not uniquely determined, then there are at least two distinct functions, each of which satisfies the conditions (17.8)–(17.10). The difference $y(x)$ of these functions must be a non-trivial solution of the equation $L_x(y) = 0$ satisfying the boundary conditions under consideration. Accordingly we obtain the following theorem.

THEOREM 18.1.   The Green's function $G(x, \xi)$ for the equation $L_x(y) = 0$ under the prescribed boundary conditions is uniquely determined, if there is no non-trivial solution $y(x) \not\equiv 0$ of the same boundary value problem and if the above construction of $C_1(\xi)$, $C_2(\xi)$, $C_3(\xi)$ and $C_4(\xi)$ is possible.[1]

Proof.   By its construction, $G(x, \xi)$, given by (18.5), evidently satisfies the conditions (17.8)–(17.10). To prove the symmetry property (17.11), that is, the property $G(x, \xi) \equiv G(\xi, x)$, let

$$y(x) = G(x, \xi_1), \quad z(x) = G(x, \xi_2) \qquad (a < \xi_1 < \xi_2 < b)$$

---

[1] We do not enter into the question of the compatibility of equations (18.5)–(18.8) to determine $C_1(\xi)$, $C_2(\xi)$, $C_3(\xi)$ and $C_4(\xi)$. Concerning this, the reader is referred to Ince, E. L., *Ordinary Differential Equations*, London, 1927; Coddington, E. A., and Levinson, N., *Theory of Ordinary Differential Equations*, New York, 1955.

We apply Green's formula to $y(x)$ and $z(x)$ on the intervals

$$(a, \xi_1 - \varepsilon), \ (\xi_1 + \varepsilon, \xi_2 - \varepsilon), \ (\xi_2 + \varepsilon, b)$$

respectively. Then, letting $\varepsilon \downarrow 0$, we obtain

$$\lim_{\varepsilon \downarrow 0} P(y(x), z(x))_{\xi_1 - \varepsilon}^{\xi_1 + \varepsilon} + \lim_{\varepsilon \downarrow 0} P(y(x), z(x))_{\xi_2 - \varepsilon}^{\xi_2 + \varepsilon} = 0$$

where

$$P(u, v) = p(x)[u(x)v'(x) - u'(x)v(x))]$$

since both $y(x)$ and $z(x)$ satisfy the boundary conditions, and $L_x(y) = 0$ for $x \neq \xi_1$ and $L_x(z) = 0$ for $x \neq \xi_2$. From the discontinuities of $y(x)$ and $z(x)$ at the points $x = \xi_1$ and $x = \xi_2$, respectively, we obtain from the above equality that

$$G(\xi_1, \xi_2) - G(\xi_2, \xi_1) = 0$$

Then, by the continuity of $G(x, \xi)$, we obtain

$$G(\xi_1, \xi_2) = G(\xi_2, \xi_1) \qquad (a \leqq \xi_1 \leqq \xi_2 \leqq b)$$

In the same way, we can also obtain the same relation for $\xi_1 \geqq \xi_2$. Thus, the proof is completed, q. e. d.

EXAMPLE. $L_x = d^2/dx^2$, $y(0) = -y(1)$, $y'(0) = -y'(1)$. Green's function $G(x, \xi)$ for this boundary value problem may be written in the form

$$G(x, \xi) = -\frac{1}{2}|x - \xi| + \frac{1}{4}$$

The general solution of $y'' = 0$ is of the form $C_1 x + C_2$; hence, if the boundary conditions are satisfied, then the solution must be identically zero. Accordingly we can apply our procedure to this problem. Now, taking as a fundamental system of the solutions of $y'' = 0$

$$Y_1(x) \equiv x, \qquad Y_2(x) \equiv 1$$

we obtain relations, corresponding to (18.6)–(18.8), as follows:

$$C_1 \xi + C_2 = C_3 \xi + C_4, \ C_1 - C_3 = 1, \ C_2 = -(C_3 + C_4), \ C_1 = -C_3$$

Solving these equations, we obtain

$$C_1 = \frac{1}{2}, \quad C_2 = \frac{1}{4} - \frac{\xi}{2}, \quad C_3 = -\frac{1}{2}, \quad C_4 = \frac{1}{4} + \frac{\xi}{2}$$

and therewith

$$G(x, \xi) = -\frac{1}{2}|x - \xi| + \frac{1}{4}$$

Generalized Green's function. Our procedure is no longer applicable when there exists a non-trivial solution $y_0(x) \not\equiv 0$ of the equation $L_x(y) = 0$ satisfying the prescribed boundary conditions. In this case, however, we may assume without loss of generality that every non-trivial solution is a constant multiple of $y_0(x)$. In fact, if there exist two non-trivial solutions which are linearly independent, then every solution of the equation satisfies the boundary conditions, because of the linear homogeneity of the conditions. Hence the boundary conditions are no longer actual restrictions. We naturally omit such a trivial eventuality.

If, in our case, there exists a solution $y(x)$ of the inhomogeneous equation $L_x(y) = -\varphi(x)$ satisfying the boundary conditions, then $\varphi(x)$ must satisfy

$$(18.9) \qquad \int_a^b \varphi(x)y_0(x)dx = 0$$

In fact, by making use of Green's formula, we obtain

$$-\int_a^b y_0(x)\varphi(x)dx$$

$$= \int_a^b \{y_0(x)L_x(y) - y(x)L_x(y_0)\}dx$$

$$= [p(x)(y_0(x)y'(x) - y_0'(x)y(x))]_a^b = 0$$

On the other hand, the solution $y(x)$ may be written in the form

$$y(x) = z(x) + Cy_0(x)$$

where $z(x)$ is a solution of $L_x(z) = -\varphi(x)$ satisfying the boundary conditions (the existence of a solution is already assumed). Since $y_0(x) \not\equiv 0$, we can choose the constant C so as

$$(18.10) \qquad \int_a^b y(x)y_0(x)dx = 0$$

Now, we shall prove that such a solution $y(x)$ of the boundary value problem satisfying (18.10) can be written as

$$(18.11) \qquad y(x) = \int_a^b G(x, \xi)\varphi(\xi)d\xi$$

by means of the *generalized Green's function* $G(x, \xi)$. By a generalized Green's function, we mean a function $G(x, \xi)$ satisfying the following five conditions: The equation (17.8),

(17.9') If $x \neq \xi$, $G(x, \xi)$ satisfies the equation $L_x(G(x, \xi)) = y_0(x)y_0(\xi)$ as a function of x. $G_x(x, \xi)$ is bounded in the region $x \neq \xi$,

the equations (17.10), (17.11), and

$$(18.12) \qquad \int_a^b G(x, \xi)y_0(x)dx = 0$$

If the generalized Green's function $G(x, \xi)$ exists, then, similarly as before, we can prove the representation (18.11), and further, we can prove that the condition (18.9) is sufficient for the existence of solutions, namely, the function $y(x)$ defined by (18.11) satisfies the equation $L_x(y) = -\varphi(x)$ and the boundary conditions. Also the solution $y(x)$ satisfies (18.10), which follows from (18.12).

The procedure of the construction of the generalized Green's function is essentially the same as in Part 18, and will therefore be omitted.

EXAMPLE 1. $L_x = d^2/dx^2$, $y'(0) = y'(1) = 0$. The general solution of $y''(x) = 0$ is a polynomial in x of degree 1. Hence, there exists a non-trivial solution $y_0(x) \equiv 1$ of the boundary value problem. By (17.9'), we have

$$L_x(G(x, \xi)) = 1, \text{ that is, } G_{xx}(x, \xi) = 1$$

Hence we have

$$G(x, \xi) = A_1 + A_2 x + \frac{x^2}{2}, \quad x \leq \xi$$

$$= B_1 + B_2 x + \frac{x^2}{2}, \quad x > \xi$$

By the boundary conditions, $G_x(0, \xi) = 0$, $G_x(1, \xi) = 0$, we obtain $A_2 = 0$, $B_2 = -1$. Then the condition

$$G_x(\xi + 0, \xi) - G_x(\xi - 0, \xi) = -1$$

holds automatically. By the continuity at $x = \xi$, that is, $G(\xi+0, \xi) - G(\xi - 0, \xi) = 0$, we obtain $B_1 - \xi - A_1 = 0$. Hence we obtain

$$G(x, \xi) = A_1 + \frac{x^2}{2}, \quad x \leq \xi$$

$$= A_1 + \xi - x + \frac{x^2}{2}, x > \xi$$

Finally, by $\int_0^1 G(x, \xi)y_0(\xi)d\xi = 0$, we obtain $A_1 = 0$. Thus the generalized Green's function is given by

$$G(x, \xi) = \frac{x^2}{2}, \qquad\qquad x \leq \xi$$

$$= \xi - x + \frac{x^2}{2}, \quad x > \xi$$

EXAMPLE 2. $L_x = d^2/dx^2, y(-1) = y(1), y'(-1) = y'(1)$. In this case,

$$y_0(x) \equiv 1/\sqrt{2}$$

is a non-trivial solution of the boundary value problem. From

$$L_x(G(x, \xi)) = \frac{1}{2}, \text{ i.e., } G_{xx}(x, \xi) = \frac{1}{2},$$

it is easily seen that $G(x, \xi)$ must be of the form

$$G(x, \xi) = \begin{cases} A_1 + A_2 x + \dfrac{1}{4} x^2, & x \leq \xi \\[2mm] B_1 + B_2 x + \dfrac{1}{4} x^2, & x > \xi \end{cases}$$

By the continuity at $x = \xi$, we obtain

$$A_1 - B_1 = (B_2 - A_2)\xi$$

and, by $G_x(\xi + 0, \xi) - G_x(\xi - 0, \xi) = -1$,

$$A_2 - B_2 = 1$$

By the boundary conditions, we obtain

$$A_1 - B_1 - (A_2 + B_2) = 0, \ A_2 - B_2 = 1$$

and hence,

$$A_1 - B_1 = -\xi, \ A_2 - B_2 = 1, \ A_2 + B_2 = -\xi$$

Therefore,

$$A_2 = \frac{1 - \xi}{2}, \ B_2 = \frac{-1 - \xi}{2}$$

Finally, by (18.12), that is,

$$\int_{-1}^{1} G(x, \xi)y_0(x)dx = 0 = \left\{\int_{-1}^{\xi} + \int_{\xi}^{1}\right\} G(x, \xi)y_0(x)dx$$

we obtain

$$A_1(\xi + 1) + A_2\left(\frac{\xi^2 - 1}{2}\right) + \frac{\xi^3 + 1}{12}$$

$$+ B_1(1 - \xi) + B_2\left(\frac{1 - \xi^2}{2}\right) + \frac{1 - \xi^3}{12} = 0$$

Since $A_1 - B_1 = -\xi$, $A_2 - B_2 = 1$,

$$A_1 + B_1 = \frac{\xi^2}{2} + \frac{1}{3}$$

Combining this with $A_1 - B_1 = -\xi$, we obtain

$$A_1 = \frac{\xi^2}{4} - \frac{\xi}{2} + \frac{1}{6}, \quad B_1 = \frac{\xi^2}{4} + \frac{\xi}{2} + \frac{1}{6}$$

Thus the generalized Green's function is given by

$$G(x, \xi) = \begin{cases} (x - \xi^2)\dfrac{1}{4} + \dfrac{x - \xi}{2} + \dfrac{1}{6}, & x \leqq \xi \\[2mm] (x - \xi)^2\dfrac{1}{4} - \dfrac{x - \xi}{2} + \dfrac{1}{6}, & x > \xi \end{cases}$$

$$= -\frac{1}{2}|x - \xi| + \frac{1}{4}(x - \xi)^2 + \frac{1}{6}$$

§ 2. Hilbert-Schmidt theory of integral equations with symmetric kernel

19. *Ascoli-Arzelà theorem*

We first consider the integral equation

$$(19.1) \qquad y(x) = \lambda \int_a^b G(x, \xi) r(\xi) y(\xi) d\xi$$

which is equivalent to a boundary value problem for $L_x(y) = -\lambda r(x)y$, as was shown in § 1. Multiply both sides of (19.1) by $(r(x))^{1/2}$. (It is assumed that $r(x) > 0$, see Part 16.) Then, setting

$$(19.2) \qquad z(x) = (r(x))^{1/2}y(x), \; K(x, \xi) = (r(x))^{1/2}G(x, \xi)(r(\xi))^{1/2}$$

we obtain

$$(19.3) \qquad z(x) = \lambda \int_a^b K(x, \xi) z(\xi) d\xi$$

Here $K(x, \xi)$ is a real-valued continuous function which has the symmetry property

$$(19.4) \qquad K(x, \xi) \equiv K(\xi, x)$$

just as $G(x, \xi)$.

If the integral equation (19.3) with the *symmetric kernel* $K(x, \xi)$ has a non-trivial continuous solution $z(x)$ for some value of parameter

λ, then, such λ is called an eigenvalue and $z(x)$ an eigenfunction belonging to the eigenvalue λ. The Hilbert-Schmidt theory for this eigenvalue problem is based upon the following existence theorem:

THEOREM 19.1. If $K(x, \xi)$ is a real-valued continuous symmetric kernel and does not vanish identically, then the integral equation (19.3) with the kernel $K(x, \xi)$ possesses at least one non-zero eigenvalue.

We shall prove, in this paragraph, the *Ascoli-Arzelà theorem*, which will play a fundamental role in the proof of Theorem 19.1.

By means of the kernel $K(x, \xi)$, every complex-valued continuous function $f(x)$ defined on the interval $[a, b]$ is transformed into a continuous function

$$\int_a^b K(x, \xi) f(\xi) d\xi$$

on the same interval. This is so because

$$\left| \int_a^b (K(x_1, \xi) - K(x_2, \xi)) f(\xi) d\xi \right|$$

$$\leq \left\{ \sup_{a \leq \xi \leq b} |f(\xi)| \right\} \int_a^b | K(x_1, \xi) - K(x_2, \xi) | d\xi$$

tends to zero as $|x_1 - x_2| \to 0$ because of the uniform continuity of $K(x, \xi)$. We denote by K the transformation

$$(19.5) \qquad (Kf)(x) = \int_a^b K(x, \xi) f(\xi) d\xi$$

Obviously the transformation K satisfies

$$(19.6) \qquad \begin{aligned} & K(f_1 + f_2) = Kf_1 + Kf_2 \\ & K(\alpha f) = \alpha(Kf), \quad \text{(for any constant } \alpha) \end{aligned}$$

Accordingly, the transformation K is a *linear operator*. For each continuous function $f(x)$, we define its *norm* by

$$(19.7) \qquad \|f\| = \left(\int_a^b |f(x)|^2 dx \right)^{1/2}$$

Then we can prove that

$$(19.8) \qquad |(Kf)(x)| \leq \|f\| \left(\int_a^b | K(x, \xi) |^2 d\xi \right)^{1/2}$$

$$(19.9) \qquad |(Kf)(x_1) - (Kf)(x_2)|$$

$$\leq \|f\| \left(\int_a^b | K(x_1, \xi) - K(x_2, \xi) |^2 d\xi \right)^{1/2}$$

In fact, these are easily derived from the *Schwarz inequality*

$$(19.10) \qquad \left| \int_a^b g(\xi)h(\xi)d\xi \right|^2 \leq \int_a^b |g(\xi)|^2 d\xi \int_a^b |h(\xi)|^2 d\xi$$

which will be proved in Part 20. Accordingly, we obtain the following theorem.

THEOREM 19.2. Let $\{f_n(x)\}$ be a sequence of continuous functions and $\|f_n\| \leq 1$, $n = 1, 2, 3, \cdots$. Let $g_n(x) = (Kf_n)(x)$. Then $\{g_n(x)\}$ satisfies

$$(19.8') \qquad \sup_{a \leq x \leq b, n \geq 1} |g_n(x)| < \infty$$

$$(19.9') \qquad \lim_{\delta \downarrow 0} \{\sup_{\delta \geq |x_1 - x_2|, n \geq 1}, \ |g_n(x_1) - g_n(x_2)|\} = 0$$

REMARK. A set of functions $\{g_n(x)\}$ is said to be *equibounded* on $[a, b]$ if it satisfies (19.8′), and *equicontinuous* on $[a, b]$ if it satisfies (19.9′). It should be noted that Theorem 19.2 does not hold for all linear operators. For example, an operator T defined by

$$(19.11) \qquad f(x) \rightarrow (Tf)(x) = \alpha(x)f(x)$$

where $\alpha(x)$ is a continuous function does not satisfy (19.8′) nor (19.9′).

Owing to the properties (19.8′) and (19.9′) of the operator K, we can apply the Ascoli-Arzelà theorem, which reads as follows:

THEOREM 19.3. Let $\{g_n(x)\}$ be a sequence of continuous functions. If $\{g_n(x)\}$ satisfies the conditions (19.8′) and (19.9′), then we can choose a subsequence $\{g_{n'}(x)\}$ which converges uniformly on the interval $[a, b]$.

Proof. Since the set of all rational numbers in the interval $[a, b]$ is denumerable, it may be arranged as

$$r_1, r_2, r_3, \cdots$$

On account of (19.8′), the sequence of numbers

$$g_1(r_1), g_2(r_1), g_3(r_1), \cdots$$

is bounded; hence by the *Bolzano-Weierstrass theorem* there exists a a convergent subsequence

$$g_{1'}(r_1), g_{2'}(r_1), g_{3'}(r_1), \cdots$$

Similarly we can select from the sequence of numbers

$$g_{1'}(r_2), g_{2'}(r_2), g_{3'}(r_2), \cdots$$

a convergent subsequence

$$g_{1\prime\prime}(r_2),\ g_{2\prime\prime}(r_2),\ g_{3\prime\prime}(r_2)\ ,\cdots$$

Repeating this procedure, we can select from each sequence of functions $(n = 1, 2, \cdots)$

$$g_1{}^{(n-1)}(x),\ g_2{}^{(n-1)}(x),\ \cdots, \qquad (g_k{}^{(0)}(x) = g_k(x))$$

a subsequence of functions

(19.12) $$g_1{}^{(n)}(x),\ g_2{}^{(n)}(x),\ g_3{}^{(n)}(x),\ \cdots \qquad (n=1, 2, 3,\cdots)$$

which converges at the points $x = r_1, r_2, \cdots, r_n$. Accordingly the subsequence

(19.13) $$g_{1\prime}(x),\ g_{2\prime\prime}(x),\ g_{3\prime\prime\prime}(x),\cdots,g_n{}^{(n)}(x),\ \cdots$$

of the original sequence $\{g_n(x)\}$ converges for every rational number $x = r_1, r_2, \cdots, r_n, \cdots$. (This method of selection (19.13) is an example of the so-called *diagonal method*.)

Next we shall prove that the sequence (19.13) converges uniformly on the interval $[a, b]$. For the sake of simplicity, we shall denote the sequence (19.13) by $\{g_n(x)\}$. According to (19.9'), there exists, for any positive number $\varepsilon > 0$, a positive number $\delta = \delta(\varepsilon) > 0$ such that

$$|g_n(x_1) - g_n(x_2)| \leqq \varepsilon \qquad \text{for all } n$$
$$\text{whenever} \quad |x_1 - x_2| \leqq \delta .$$

On the other hand, the set of all rational numbers is dense in the interval $[a, b]$, hence for the above δ, there exists a number $N = N(\delta)$ such that

$$\min_{1 \leqq k \leqq N} |x - r_k| < \delta$$

for every number x in the interval. Further, since $\{g_n(x)\}$ converges at the points $x = r_1, r_2, \cdots, r_N$, there exists, for the $\varepsilon > 0$, a number $M = M(\varepsilon)$ such that

$$m, n \geqq M \text{ implies } |g_n(r_k) - g_m(r_k)| < \varepsilon \qquad (1 \leqq k \leqq N).$$

Therefore, for each x, there exists a rational number r_k $(1 \leqq k \leqq N)$ such that

$$|g_n(x) - g_m(x)| \leqq |g_n(x) - g_n(r_k)|$$
$$+ |g_n(r_k) - g_m(r_k)| + |g_m(r_k) - g_m(x)| \leqq 3\varepsilon$$

wherever $m, n \geqq M$. This means that $\{g_n(x)\}$ converges uniformly on the interval $[a, b]$, q.e.d.

20. *Existence proof for the eigenvalues*

Inner product. For every pair of complex-valued continuous functions $f(x)$ and $g(x)$, the integral

$$(20.1) \qquad (f, g) = \int_a^b f(x)\overline{g(x)}dx$$

exists and is called the *inner product* (or *scalar product*) of the two functions $f(x)$ and $g(x)$. Obviously, the inner product has the following properties:

$$(20.2) \qquad ((\alpha f + \beta g), h) = \alpha(f, g) + \beta(g, h)$$

$$(20.3) \qquad (g, f) = \overline{(f, g)}$$

$$(20.4) \qquad \|f\|^2 = (f, f)$$

where α and β are arbitrary complex numbers. Furthermore the Schwarz inequality holds [(19.10)], that is

$$(20.5) \qquad |(g, h)| \leq \|g\| \, \|h\|$$

To prove this, let α and β be two real numbers. Then

$$\| \alpha g + \beta h \|^2 = ((\alpha g + \beta h), (\alpha g + \beta h))$$
$$= \alpha^2 \|g\|^2 + 2\alpha\beta \mathscr{R}(g, h) + \beta^2 \|h\|^2 \geq 0$$

$\mathscr{R}(g, h)$ denoting the real part of (g, h).

The condition of positivity of this quadratic form is

$$\{\mathscr{R}(g, h)\}^2 \leq \|g\|^2 \|h\|^2$$

Taking θ such that $\exp((-1)^{1/2}\theta)(g, h) = |(g, h)|$, and substituting $\exp((-1)^{1/2}\theta)g$ for g, we obtain the inequality (20.5). We also obtain

$$\|g + h\|^2 = \|g\|^2 + 2\mathscr{R}(g, h) + \|h\|^2 \leq (\|g\| + \|h\|)^2$$

and hence the *Minkowski inequality*

$$(20.6) \qquad \|g + h\| \leq \|g\| + \|h\|$$

Let $K(x, \xi)$ be a complex-valued continuous function of $x, \xi \ (a \leq x, \xi \leq b)$ which is *Hermitian symmetric*, that is,

$$(20.7) \qquad K(x, \xi) \equiv \overline{K(\xi, x)}$$

(The real-valued kernel satisfying (19.4) is of course Hermitian symmetric.) Then we have, for every pair of continuous functions $f(x)$ and $g(x)$,

$$(20.8) \qquad (Kf, g) = (f, Kg), (Kf)(x) = \int_a^b K(x, \xi) f(\xi) d\xi$$

According to (19.8), there exists a positive number α such that

$$(20.9) \qquad \|Kf\| \leqq \alpha \|f\|$$

holds for all continuous functions f. Moreover, Theorem 19.2 is valid for the operator K.

We shall now prove the existence of the eigenvalues of the operator K. We begin with

THEOREM 20.1. Let

$$(20.10) \qquad \|K\| = \sup_{\|f\|=1} \|Kf\|$$

Then obviously, by the linearity of the operator K, $\|K\|$ is the infimum of all α such that (20.9) holds for all f. Furthermore,

$$(20.11) \qquad \|K\| = \sup_{\|f\|=1} |(Kf, f)|$$

Proof. The equations (20.5) and (20.10) imply that $|(Kf, f)| \leqq \|Kf\| \|f\| \leqq \|K\|$ for every f with $\|f\| = 1$. Hence, by setting

$$\beta = \sup_{\|f\|=1} |(Kf, f)|$$

we obtain $\beta \leqq \|K\|$.

Next we shall prove $\beta \geqq \|K\|$. By virtue of the linearity of the operator K, we see that $|(Kf, f)| \leqq \beta \|f\|^2$ holds for every f without the assumption $\|f\| = 1$. Hence, taking account of (20.2), (20.3), and (20.8), we have, for two functions f, g with $\|f\| = \|g\| = 1$,

$$(K(f + g), (f + g)) = (Kf, f) + (Kg, g) + (Kf, g) + (Kg, f)$$
$$= (Kf, f) + (Kg, g) + 2\mathscr{R}(Kf, g) \leqq \beta \|f + g\|^2$$
$$(K(f - g), (f - g)) = (Kf, f) + (Kg, g) - 2\mathscr{R}(Kf, g)$$
$$\geqq -\beta \|f - g\|^2$$

Accordingly we obtain

$$4\mathscr{R}(Kf, g) \leqq \beta(\|f + g\|^2 + \|f - g\|^2)$$
$$= 2\beta(\|f\|^2 + \|g\|^2) = 4\beta$$

Therefore, if $(Kf)(x) \not\equiv 0$, we obtain $\|Kf\| \leqq \beta$ by setting $g = Kf/\|Kf\|$. In the case $(Kf)(x) \equiv 0$, we have $0 = \|Kf\| \leqq \beta \|f\|$ since $\beta \geqq 0$. Thus, in any case, we have $\beta \geqq \|K\|$ and hence the proof is completed.

We shall next prove the following theorem.

THEOREM 20.2. If $\|K\| \neq 0$, then either the inverse of $\|K\| = \beta$ or of $-\|K\| = -\beta$ is an eigenvalue of the operator K.

REMARK. If $K(x, \xi) \not\equiv 0$, then $\int_a^b |K(x_0, \xi)|^2 d\xi > 0$ for some x_0. Hence, setting $f_0(x) = K(x, x_0)$, we obtain $(Kf_0)(x_0) \neq 0$ which means $\|K\| \neq 0$. Accordingly this theorem contains Theorem 19.1.

Proof. (Kf, f) is real by (20.3) and (20.8). Hence, by (20.11), there exists a sequence $\{f_m\}$ of functions which satisfies

(20.12) $\|f_m\| = 1 (m = 1, 2, \cdots)$ and
$\lim_{m \to \infty} (Kf_m, f_m) = \beta$ (or $\lim_{m \to \infty} (Kf_m, f_m) = -\beta$)

Then, by Theorem 19.3, the sequence $\{(Kf_m)(x)\}$ contains a subsequence which converges to a continuous function $\varphi(x)$ uniformly on the interval $[a, b]$. We may assume, without loss of generality, that the sequence $\{(Kf_m)(x)\}$ itself converges uniformly to the function $\varphi(x)$. By (20.8), we have

(20.13) $\|Kf_m - \beta f_m\|^2 = \|Kf_m\|^2 + \beta^2 \|f_m\|^2 - 2\beta(Kf_m, f_m)$

On the other hand, on account of (20.12), and because $\lim_{m \to \infty} (Kf_m)(x) = \varphi(x)$, the right side of (20.13) tends to

$$\|\varphi\|^2 + \beta^2 - 2\beta^2 = \|\varphi\|^2 - \beta^2$$

as $m \to \infty$. Therefore we have $\|\varphi\| \geq \beta$ so that $\varphi(x) \not\equiv 0$.

Since $\|Kf_m\| \leq \beta \|f_m\| = \beta$, we obtain, from (20.13),

$$\lim_{m \to \infty} \|Kf_m - \beta f_m\|^2 \leq \beta^2 + \beta^2 - 2\beta^2 = 0$$

Thus $\lim_{m \to \infty} \|Kf_m - \beta f_m\| = 0$. Then (20.9) implies

$$\lim_{m \to \infty} \|K(Kf_m) - \beta Kf_m\| = 0$$

Accordingly, we obtain

$$\int_a^b |(K\varphi)(x) - \beta\varphi(x)|^2 dx = 0$$

Since $(K\varphi)(x)$ and $\varphi(x)$ are continuous, we obtain further

$$(K\varphi)(x) \equiv \beta\varphi(x)$$

In the case $\lim_{m \to \infty} (Kf_m, f_m) = -\beta$, applying the same arguments to the operator $-K$, instead of K, we see that β^{-1} is an eigenvalue of $-K$. Hence, in this case, $-\beta^{-1}$ is an eigenvalue of K.

The method of the above proof is called *"maximal method"*

because of the fact that the inverse of $\sup_{||f||=1} |(Kf, f)| = \beta$ or $-\beta$ is an eigenvalue of K.

21. The Bessel inequality. The Hilbert-Schmidt expansion theorem

Let $f(x)$ and $g(x)$ be continuous functions on the interval $[a, b]$. When

$$(21.1) \qquad\qquad (f, g) = 0$$

$f(x)$ is said to be *orthogonal* to $g(x)$, and this fact is indicated by writing $f \perp g$. Clearly, the orthogonality relation is reflexive, that is, if $f \perp g$, then $g \perp f$.

For an operator K defined by means of a symmetric (or Hermitian symmetric) kernel $K(x, \xi)$, the following theorem holds.

THEOREM 21.1. (i) All eigenvalues of the operator K are real. (ii) Two eigenfunctions, corresponding to different eigenvalues, are orthogonal.

Proof. Let λ be an eigenvalue of the operator K and φ the corresponding eigenfunction, that is, $K\varphi = \lambda^{-1}\varphi$. Then

$$(K\varphi, \varphi) = (\lambda^{-1}\varphi, \varphi) = \lambda^{-1}(\varphi, \varphi)$$

By (20.8) and (20.3), we obtain that

$$(K\varphi, \varphi) = (\varphi, K\varphi) = \overline{(K\varphi, \varphi)}$$

and hence $(K\varphi, \varphi)$ is real. On the other hand, $(\varphi, \varphi) > 0$ is real. Therefore λ must be real. This proves (i).

Let $K\varphi = \lambda^{-1}\varphi$, and $K\psi = \mu^{-1}\psi, (\lambda \neq \mu)$. Then

$$(\varphi, \psi) = (\lambda K\varphi, \psi) = \lambda(K\varphi, \psi) = \lambda(\varphi, K\psi) = \lambda(\varphi, \mu^{-1}\psi)$$
$$= \lambda\mu^{-1}(\varphi, \psi)$$

since μ is real by (i). Hence we obtain, by $\lambda\mu^{-1} \neq 1$, that $(\varphi, \psi) = 0$. This proves (ii), q.e.d.

We shall next describe *Schmidt's orthogonalization process.*

THEOREM 21.2. Let $f_1(x), f_2(x), \cdots, f_n(x), \cdots$ be linearly independent, viz., any finite number of $f_n(x)$ be linearly independent. Let

$$g_1(x) = f_1(x)/||f_1||$$
$$g_2(x) = (f_2(x) - (f_2, g_1)g_1(x))/||f_2 - (f_2, g_1)g_1||$$
$$\cdots\cdots\cdots\cdots\cdots\cdots$$
$$g_n(x) = \left(f_n(x) - \sum_{k=1}^{n-1}(f_n, g_k)g_k(x)\right)\Big/\left\|f_n - \sum_{k=1}^{n-1}(f_n, g_k)g_k\right\|$$
$$\cdots\cdots\cdots\cdots\cdots\cdots$$

Then, $g_1(x), g_2(x), \cdots, g_n(x), \cdots$ satisfy the *orthonormality relations*

$$(21.2) \qquad (g_k, g_s) = \delta_{ks} \begin{cases} = 1 & (k = s) \\ = 0 & (k \neq s) \end{cases}$$

Proof. Since f_1 and f_2 are linearly independent, $f_2 - (f_2, g_1)g_1$ is not identically zero; hence we can divide it by its norm, and thus $g_2(x)$ is well defined. Since g_2 is a linear combination of f_2 and g_1, f_3, g_1, g_2 are linearly independent. Hence $f_3 - (f_3, g_2)g_2 - (f_3, g_1)g_1$ is not identically zero; so we can divide it by its norm and g_3 is well defined. In this way, we can define $g_4, g_5, \cdots, g_n, \cdots$ successively.

Clearly g_n is *normalized*, that is, $\| g_n \| = 1$. Next we shall prove that $\{g_n\}$ satisfies (21.2). By the definition of g_2, obviously $(g_2, g_1) = 0$. Thus, by

$$(\{f_3 - (f_3, g_1)g_1 - (f_3, g_2)g_2\}, g_1) = (f_3, g_1) - (f_3, g_1) - 0 = 0$$

we have $(g_3, g_1) = 0$. In this way we see that

$$(g_3, g_1) = 0, (g_4, g_1) = 0, \cdots, (g_n, g_1) = 0, \cdots$$

Similarly, we see, starting with $(g_3, g_2) = 0$, that

$$(g_4, g_2) = 0, (g_5, g_2) = 0, \cdots, (g_n, g_2) = 0, \cdots$$

Repeating the same procedure, we finally obtain that $(g_k, g_s) = 0$ for $k \neq s$,   q.e.d.

As a by-product of the proof, we obtain the following

COROLLARY. Let $f_1, f_2, \cdots, f_n, \cdots$ and $g_1, g_2, \cdots, g_n, \cdots$ be as in Theorem 21.2. Then

(21.3) every g_n can be written as a linear combination of
$f_1, f_2, \cdots, f_n$, and moreover, every f_n can be written as
a linear combination of $g_1, g_2, \cdots, g_n$.

REMARK. A set $\{\varphi_n\}$ of functions is called an *orthonormal system* if it satisfies the orthonormality relations (21.2).

THEOREM 21.3. (i) The operator K has at most denumerably many eigenvalues. The set of all eigenvalues has no limiting point except $\pm\infty$. (ii) The *multiplicity* of every eigenvalue λ is finite. In other words, the number of linearly independent eigenfunctions corresponding to the eigenvalue λ is finite.

Proof. It should be noted first that if $K\varphi = \lambda^{-1}\varphi$ and $K\psi = \lambda^{-1}\psi$, then, by the linearity of the operator K, $K(\alpha\varphi + \beta\psi) = \alpha K\varphi + \beta K\psi =$

$\lambda^{-1}(\alpha\varphi + \beta\psi)$, which means that any linear combination of eigenfunctions corresponding to the same eigenvalue is either an eigenfunction corresponding to the same eigenvalue or identically zero. By virtue of this property, for any eigenvalue λ, there exists a set of eigenfunctions, corresponding to the eigenvalue λ, such that they are linearly independent and every eigenfunction corresponding to this eigenvalue is written as their linear combination. Suppose now such a set of eigenfunctions contains denumerably many functions f_k. Then, by Schmidt's orthogonalization process, there exists an orthonormal system $\{g_k\}$ which satisfies (21.3). Since $Kg_k = \lambda^{-1}g_k$, we have

$$K(g_j - g_k) = \lambda^{-1}g_j - \lambda^{-1}g_k , \quad (g_j, g_k) = \delta_{jk}$$

Hence, if $j \neq k$,

$$|| K(g_j - g_k) ||^2 = \lambda^{-2} + \lambda^{-2} = 2\lambda^{-2}$$

while

$$|| g_j - g_k ||^2 = 2$$

This contradicts Theorems 19.2 and 19.3. Thus (ii) is proved.

Next we shall prove (i). By Theorem 21.1, the eigenfunctions corresponding to different eigenvalues are orthogonal. Hence, if there exists a finite limiting point of eigenvalues, we would have an orthonormal system $\{\varphi_j\}$ satisfying

$$K\varphi_j = \lambda_j^{-1}\varphi_j \quad (j = 1, 2, 3, \cdots)$$
$$\lambda_j^{-1} \to \lambda^{-1}(\neq 0)$$

Then, since

$$K(\varphi_j - \varphi_k) = \lambda_j^{-1}\varphi_j - \lambda_k^{-1}\varphi_k , \quad (\varphi_j, \varphi_k) = \delta_{jk}$$

we obtain that, for sufficiently large j and $k, j \neq k$,

$$|| K(\varphi_j - \varphi_k) ||^2 = \lambda_j^{-2} + \lambda_k^{-2} \geq \lambda^{-2} , \quad || \varphi_j - \varphi_k ||^2 = 2$$

This contradicts Theorems 19.2 and 19.3. Therefore the number of eigenvalues λ satisfying $n < |\lambda| \leq n + 1$ is finite $(n = 0, 1, 2, \cdots)$. Hence, we see that the set of all eigenvalues consists of at most denumerably many points and has no limiting points except for $\pm\infty$, q.e.d.

According to Theorem 21.3, if the operator K has infinitely many eigenvalues,[1] then we can write its eigenvalues and eigenfunctions as

(21.4)
$$|\lambda_1| \leqq |\lambda_2| \leqq \cdots, \lim_{j \to \infty} |\lambda_j| = \infty$$
$$K\varphi_j = \lambda_j^{-1}\varphi_j \quad (j = 1, 2, \cdots), \quad (\varphi_j, \varphi_k) = \delta_{jk}$$

so that every eigenvalue λ of K is equal to some λ_j and every eigenfunction corresponding to the eigenvalue λ is written as a linear combination of finitely many eigenfunctions φ_j corresponding to the eigenvalues $\lambda_j = \lambda$. We shall refer to $\{\varphi_j\}$ as the *complete orthonomal system of eigenfunctions* of the operator K (or the kernel K).

REMARK. As was shown in §1 of Chapter 2, the eigenvalue problem for the boundary value problem of Sturm-Liouville type is equivalent to that for the symmetric kernel. Hence Theorem 21.3 holds true for the eigenvalues of our boundary value problem.

Bessel inequality. Let $f(x)$ be a continuous function and let $\{\varphi_j\}$ be the orthonormal system as in (21.2). The numbers

(21.5)
$$(f, \varphi_j) \quad (j = 1, 2, \cdots)$$

are called the *Fourier coefficients* of $f(x)$ with respect to the orthonormal system $\{\varphi_j\}$. In terms of these coefficients we have the *Bessel inequality*:

(21.6)
$$||f||^2 \geqq \sum_{j=1}^{\infty} |(f, \varphi_j)|^2$$

Proof. Put $g_n = f - \sum_{j=1}^{n}(f, \varphi_j)\varphi_j$. Then, from the orthonormality of the system $\{\varphi_j\}$, it follows that

$$0 \leqq ||g_n||^2 = ||f||^2 - \sum_{j=1}^{n} |(f, \varphi_j)|^2, \quad \text{q.e.d.}$$

Next we shall prove the Hilbert-Schmidt expansion theorem:

THEOREM 21.4. Let $f(x)$ be a continuous function defined on the interval $[a, b]$. Let

(21.7)
$$(Kf)(x)$$

be its transform by the operator K, and

(21.8)
$$\sum_{j=1}^{\infty} (Kf, \varphi_j)\varphi_j(x)$$

the *Fourier expansion* of (21.7) with respect to the complete orthonormal system of eigenfunctions $\{\varphi_j\}$. Then (21.8) converges absolutely to (21.7), uniformly on the interval $[a, b]$.

---

[1] Kernels do not necessarily have infinitely many eigenvalues. See the remark at the end of this section.

Proof. The equation (21.4) yields

$$(Kf, \varphi_j) = (f, K\varphi_j) = (f, \lambda_j^{-1}\varphi_j) = \lambda_j^{-1}(f, \varphi_j)$$

Applying the Bessel inequality (21.6) to $K(x, \xi)$ as a function of ξ, we obtain

$$\sum_{j=n}^{m} \left| \int_a^b K(x, \xi)\varphi_j(\xi)d\xi \right|^2 = \sum_{j=n}^{m} \lambda_j^{-2} |\varphi_j(x)|^2 \leqq \int_a^b |K(x, \xi)|^2 d\xi$$

Therefore, by Schwarz inequality,

$$\sum_{j=n}^{m} |(Kf, \varphi_j)| \, |\varphi_j(x)| = \sum_{j=n}^{m} |(f, \varphi_j)| \, |\lambda_j^{-1}\varphi_j(x)|$$

$$\leqq \left\{ \sum_{j=n}^{m} |(f, \varphi_j)|^2 \right\}^{1/2} \left\{ \int_a^b |K(x, \xi)|^2 d\xi \right\}^{1/2}$$

According to the Bessel inequality (21.6), the first term on the right side tends to zero as $n, m \to \infty$. Furthermore, since $K(x, \xi)$ is continuous for $a \leqq x \leqq b, a \leqq \xi \leqq b$, the second term is a continuous function on the closed interval $[a, b]$, and hence is bounded. Hence (21.8) converges absolutely and uniformly on $[a, b]$.

We shall next prove that (21.8) is equal to (21.7). To do this, put

(21.9) $$K_n(x, \xi) = K(x, \xi) - \sum_{j=1}^{n} \lambda_j^{-1}\varphi_j(x)\overline{\varphi_j(\xi)}$$

Then,

(21.10) $$K_n(x, \xi) = \overline{K_n(\xi, x)}$$

since every eigenvalue λ_j is real. Let us define the operator K_n by means of the function $K_n(x, \xi)$, in the same manner as the operator K. Then we obtain

$$K_n f = Kf - \sum_{j=1}^{n} \lambda_j^{-1}(f, \varphi_j)\varphi_j = Kf - \sum_{j=1}^{n} (f, \lambda_j^{-1}\varphi_j)\varphi_j$$

$$= Kf - \sum_{j=1}^{n} (f, K\varphi_j)\varphi_j = Kf - \sum_{j=1}^{n} (Kf, \varphi_j)\varphi_j$$

Hence it is sufficient to prove that $(K_n f)(x)$ converges uniformly to zero. Since the system $\{\varphi_j\}$ is orthonormal, we have

$$(K_n f, \varphi_m) = (Kf, \varphi_m) - (Kf, \varphi_m) = 0$$

for $m \leqq n$. From this, it follows that, if f is an eigenfunction of the operator K_n, corresponding to an eigenvalue δ, then

$$(f, \varphi_m) = 0 \quad (m \leqq n)$$

and hence, by (21.9), $Kf = K_n f = \delta^{-1}f$; this means that δ is also an eigenvalue of K and the corresponding eigenfunction is f. Accordingly, f must be written as a linear combination of eigenfunctions φ_j corresponding to the eigenvalues $\lambda_j = \delta$, namely,

$$f = \sum_{\lambda_j = \delta} \beta_j \varphi_j , \qquad \beta_j = (f, \varphi_j)$$

from which we derive that, for $j < n$, $\beta_j = 0$. Hence we see that every eigenvalue δ of the operator K_n satisfies

$$| \delta | \geqq | \lambda_n |$$

Applying the maximal method as in Theorem 20.2 to the operator K_n, we obtain

$$\| K_n \| \leqq | \lambda_n |^{-1}$$

and hence $\lim_{n \to \infty} \| K_n \| = 0$, since $\lim_{n \to \infty} | \lambda_n | = \infty$. Thus, we obtain that $\lim_{n \to \infty} \| K_n f \| = 0$. Hence we obtain that the uniform limit $\lim_{n \to \infty} (K_n f)(x)$, which as was already shown does exist, satisfies

$$(21.11) \qquad \| \lim_{n \to \infty} K_n f \| = 0$$

On the other hand, $(Kf)(x)$ is, together with $f(x)$, continuous, and $\varphi_n(x) = \lambda_n \int_a^b K(x, \xi) \varphi_n(\xi) d\xi$ is, together with $K(x, \xi)$, continuous; hence $(K_n f)(x)$ is continuous, so that its uniform limit $\lim_{n \to \infty} (K_n f)(x)$ is also continuous. Therefore, by (21.11), we obtain that $\lim_{n \to \infty} (K_n f)(x) \equiv 0$, q.e.d.

Expansion theorem and Fourier series expansion. If, for a given continuous function $\psi(x)$, $y(x)$ satisfies the inhomogeneous equation

$$(21.12) \qquad (p(x)y'(x))' - q(x)y(x) = - \psi(x)$$

and the prescribed boundary conditions, then, as was already shown in Parts 16–18, $y(x)$ can be written as

$$(21.13) \qquad y(x) = \int_a^b G(x, \xi) \psi(\xi) d\xi$$

by means of the Green's function $G(x, \xi)$ for the associated homogeneous equation with the boundary conditions under consideration, whenever the Green's function exists. It is also shown in Part 18 that even when the Green's function does not exist, namely, when there exists a non-trivial solution $y_0(x)$ of the boundary value problem under consideration, $y(x)$ is written as (21.13) by means of the generalized Green's function $G(x, \xi)$ if $y(x)$ satisfies the additional condition

$$(21.14) \qquad \int_a^b y(x)y_0(x)dx = 0$$

Accordingly, by the expansion theorem (Theorem 21.4), $y(x)$ can be expanded in a Fourier series, which converges absolutely and uniformly on the interval $[a, b]$, with respect to the complete orthonormal system $\{\varphi(x)\}$ of solutions of the equation

$$(21.15) \qquad (p(x)\varphi'(x))' - q(x)\varphi(x) = -\lambda\varphi(x) \quad (\lambda \neq 0)$$

satisfying the boundary conditions under consideration.

EXAMPLE 1. Let $y(x)$ be a twice continuously differentiable function on the interval $[0, 1]$ and $y(0) = y(1) = 0$. Then $y(x)$ satisfies

$$y''(x) = -\psi(x), \text{ where } \psi(x) = -y''(x).$$

Accordingly, $y(x)$ can be expanded in the Fourier series

$$y(x) = \sum_{n=1}^{\infty} \alpha_n \sqrt{2} \cdot \sin(n\pi x)$$

$$\alpha_n = \int_0^1 y(x) \sqrt{2} \cdot \sin(n\pi x)dx$$

which converges uniformly on the interval $[0, 1]$, with respect to the orthonormal system $\{\sqrt{2} \cdot \sin(n\pi x)\}$ $(n = 1, 2, \cdots)$ of the solutions of the boundary value problem

$$\varphi''(x) = -\lambda\varphi(x), \qquad \varphi(0) = \varphi(1) = 0$$

EXAMPLE 2. Let $y(x)$ be a twice continuously differentiable function on the interval $[0, 1]$ which satisfies

$$\int_0^1 y(x)dx = 0, \ y(0) = y(1), \ y'(0) = y'(1)$$

Then $y(x)$ can be expanded in the Fourier series

$$y(x) = \sum_{n=1}^{\infty} \{\alpha_n \sqrt{2} \cdot \sin(2n\pi x) + \beta_n \sqrt{2} \cdot \cos(2n\pi x)\}$$

$$\alpha_n = \sqrt{2} \int_0^1 y(x) \sin(2n\pi x)dx$$

$$\beta_n = \sqrt{2} \int_0^1 y(x) \cos(2n\pi x)dx$$

which converges uniformly on the interval $[0, 1]$, with respect to the orthonormal system $\{\sqrt{2} \cdot \sin(2n\pi x), \sqrt{2} \cdot \cos(2n\pi x)\}$ $(n = 1, 2, \cdots)$ of the solutions of the (periodic) boundary value problem

$$\varphi''(x) = -\lambda\varphi(x), \quad \varphi(0) = \varphi(1), \quad \varphi'(0) = \varphi'(1)$$

If $y(x)$ satisfies only the conditions for the periodicity, that is, $y(0) = y(1)$, $y'(0) = y'(1)$, then, applying the above result to the function

$$y(x) - \int_0^1 y(x)dx$$

we obtain that every twice continuously differentiable periodic function $y(x)$ with period 1 can be expanded in the Fourier series, which converges uniformly, with respect to the orthogonal system

$$\{\sqrt{2} \cdot \sin(2n\pi x), \sqrt{2} \cdot \cos(2n\pi x)\}, \qquad (n = 0, 1, 2, \cdots)$$

This is known as *Fourier series expansion*.

REMARK. According to Theorem 20.2, the operator K has at least one eigenvalue, whenever $K(x, \xi)$ is not identically zero. In certain cases, the operator K has only a finite number of eigenvalues, although $K(x, \xi) \neq 0$. For example, consider an operator K defined by the kernel $K(x, \xi) = 2e^{x+\xi}$, or, the equation

$$\varphi(x) - \lambda \int_0^1 2e^{x+\xi}\varphi(\xi)d\xi = 0$$

Setting

$$\int_0^1 e^\xi \varphi(\xi)d\xi = C$$

and substituting this in the given equation, we see that

$$\varphi(x) = 2\lambda C e^x$$

Substituting this once more in the given equation, we obtain

$$\lambda = \frac{1}{e^2 - 1}, \quad \varphi(x) = \frac{2C}{e^2 - 1}e^x$$

Hence the operator K has only one eigenvalue with multiplicity one. A Hermitian symmetric kernel $K(x, \xi)$ which is of the form

$$K(x, \xi) = \sum_{i=1}^n a_i(x)\overline{a_i(\xi)}$$

is called a *degenerated kernel*. A degenerated kernel has only a finite number of eigenvalues. To prove this, let

$$\varphi(x) = \lambda \int_a^b K(x, \xi)\varphi(\xi)d\xi$$

and set

$$C_i = \int_a^b \overline{a_i(\xi)}\varphi(\xi)d\xi$$

Then, similarly as above, it is easy to see that λ must be a root of an algebraic equation of degree n. Hence the number of all eigen-values in question is at most n.

This is an example of an integral equation which can be reduced to an algebraic equation. We shall show later (Part 27–28) that the fundamental theorem of the Fredholm integral equation is derived by a similar procedure.

22. Approximations of eigenvalues. Rayleigh's principle and the Kryloff-Weinstein theorem

Let $\{\varphi_n\}$ be an orthonormal system, and let

$$(22.1) \qquad y(x) = \sum_{n=1}^{\infty} c_n\varphi_n(x), \ z(x) = \sum_{n=1}^{\infty} d_n\varphi_n(x)$$

be two Fourier expansions which converge absolutely and uniformly on the interval $[a, b]$. Then, integrating term by term, we obtain

$$(22.2) \qquad (y, z) = \sum_{n=1}^{\infty} c_n\bar{d}_n$$

on account of the orthonormality of $\{\varphi_n\}$. In particular, setting $z = y$, we have

$$(22.3) \qquad \| y \|^2 = \sum_{n=1}^{\infty} | c_n |^2$$

which is known as the *Parseval completeness relation*. As is easily seen from the proof of (21.6), the equation (22.3) means that

$$(22.4) \qquad \lim_{n\to\infty} \| y - \sum_{j=1}^{n} c_j\varphi_j \|^2 = 0$$

that is,

$$(22.4') \qquad \lim_{n\to\infty} \int_a^b | y(x) - \sum_{j=1}^{n} c_j\varphi_j(x) |^2 dx = 0$$

Hence (22.3) is a natural consequence of (22.4).

By making use of the completeness relation, we can obtain the successive approximations of the eigenvalue as follows:

THEOREM 22.1. Suppose that every eigenvalue of the operator K is positive. Let $g_0(x) \not\equiv 0, g_1(x) = (Kg_0)(x) \not\equiv 0$,

$$(22.5) \qquad g_2 = Kg_1 = K^2g_0, g_3 = Kg_2 = K^2g_1 = K^3g_0, \cdots$$

and

$$(22.6) \qquad \beta_n = \frac{\| g_n \|}{\| g_{n+1} \|}, \qquad \alpha_n = \frac{(g_{n+1}, g_n)}{\| g_{n+1} \|^2}$$

Then

$$(22.7) \qquad\qquad 0 < \alpha_n \leqq \beta_n$$

and $\{\beta_n\}$ is monotone decreasing. Further, there exists an eigen-value λ of K such that $\{\alpha_n\}$ and $\{\beta_n\}$ tend to λ, as $n \to \infty$, and $\alpha_n \geqq \lambda$.

Proof. By the expansion theorem, every g_n can be expanded as follows:

$$(22.8) \quad g_n = K^{n-1}g_1 = \sum_{j=1}^{\infty} (K^{n-1}g_1, \varphi_j)\varphi_j = \sum_{j=1}^{\infty} (g_1, K^{n-1}\varphi_j)\varphi_j$$

$$= \sum_{j=1}^{\infty} (g_1, \lambda_j^{-(n-1)}\varphi_j)\varphi_j = \sum_{j=j_0}^{\infty} (g_1, \varphi_j)\lambda_j^{-(n-1)}\varphi_j$$

where j_0 is the least j for which $(g_1, \varphi_j) \neq 0$. Such j_0 surely exists, for if otherwise, we would have, by Theorem 21.4,

$$(22.9) \qquad\qquad g_1(x) = \sum_{j=1}^{\infty} (g_1, \varphi_j)\varphi_j(x) \equiv 0$$

which contradicts the assumption $g_1(x) = (Kg_0)(x) \neq 0$. The equation (22.8) together with (22.2) implies that

$$(22.10) \qquad (g_{n+1}, g_n) = \sum_{j=j_0}^{\infty} \lambda_j^{-2n+1} |(g_1, \varphi_j)|^2$$

$$||g_n||^2 = \sum_{j=j_0}^{\infty} \lambda_j^{-2(n-1)} |(g_1, \varphi_j)|^2$$

Hence, by the assumption

$$(22.11) \qquad\qquad 0 < \lambda_1 \leqq \lambda_2 \leqq \lambda_3 \leqq \cdots$$

we obtain that

$$\lambda_{j_0} \leqq \alpha_n = \lambda_{j_0} \left\{ \frac{|(g_1, \varphi_{j_0})|^2 + \sum_{j=j_0+1}^{\infty} (\lambda_j \lambda_{j_0}^{-1})^{-2n+1} |(g_1, \varphi_j)|^2}{|(g_1, \varphi_{j_0})|^2 + \sum_{j=j_0+1}^{\infty} (\lambda_j \lambda_{j_0}^{-1})^{-2n} |(g_1, \varphi_j)|^2} \right\}$$

Furthermore, making use of the Schwarz inequality (20.5), we see that $||g_{n+1}|| \, ||g_n|| \geqq |(g_{n+1}, g_n)|$. Thus we obtain $\alpha_n \leqq \beta_n$.

Next we shall prove that $\{\beta_n\}$ is monotone decreasing and tends to λ_{j_0} as $n \to \infty$. Using the Schwarz inequality again we have

$$|(g_{n+1}, g_{n-1})| \leqq ||g_{n+1}|| \, ||g_{n-1}||$$

And since

$$|(g_{n+1}, g_{n-1})| = |(Kg_n, g_{n-1})| = |(g_n, Kg_{n-1})| = ||g_n||^2$$

we obtain that

$$||g_n|| / ||g_{n-1}|| \leqq ||g_{n+1}|| / ||g_n||$$

This means that $\{\beta_n\}$ is monotone decreasing; hence $\lim_{n \to \infty} \beta_n$ does exist. For the sequence $\{\|g_n\|\}$ of positive numbers, the following inequality holds:

(22.12)
$$\underline{\lim}_{n\to\infty}\frac{\|g_n\|}{\|g_{n+1}\|} \leq \underline{\lim}_{n\to\infty}\|g_n\|^{-(1/n)}$$

$$\leq \overline{\lim}_{n\to\infty}\|g_n\|^{-(1/n)} \leq \overline{\lim}_{n\to\infty}\frac{\|g_n\|}{\|g_{n+1}\|}$$

Thus, in order to prove that $\lim_{n\to\infty}\beta_n = \lambda_{j_0}$, it is sufficient to show that

(22.13)
$$\lim_{n\to\infty}{}^{2(n+1)}\!\sqrt{(\|g_{n+1}\|^2)} = \lambda_{j_0}^{-1}$$

holds true.

Proof of (22.13). The equation (22.10) implies that

(22.14)
$${}^{2n}\!\sqrt{(\|g_{n+1}\|^2)} = \lambda_{j_0}^{-1}\,{}^{2n}\!\sqrt{\left[\sum_{j=j_0}^{\infty}\left(\frac{\lambda_{j_0}}{\lambda_j}\right)^{2n}|(g_1,\varphi_j)|^2\right]}$$

Hence, by (22.11) and (22.10), we obtain

$$\overline{\lim}_{n\to\infty}{}^{2n}\!\sqrt{(\|g_{n+1}\|^2)} \leq \lambda_{j_0}^{-1}\overline{\lim}_{n\to\infty}{}^{2n}\!\sqrt{(\|g_1\|^2)} = \lambda_{j_0}^{-1}$$

On the other hand, from (22.14) there follows, by virtue of $(g_1,\varphi_{j_0})\neq0$, that

$$\underline{\lim}_{n\to\infty}{}^{2n}\!\sqrt{(\|g_{n+1}\|^2)} \geq \lambda_{j_0}^{-1}\underline{\lim}_{n\to\infty}{}^{2n}\!\sqrt{[|(g_1,\varphi_{j_0})|^2]} \geq \lambda_{j_0}^{-1}$$

Thus we obtain

$$\lim_{n\to\infty}{}^{2n}\!\sqrt{(\|g_{n+1}\|^2)} = \lambda_{j_0}^{-1}$$

This completes the proof of the theorem, q.e.d.

REMARK 1. For the case when every eigenvalue of K is negative, we may replace α_n by $|(g_{n+1},g_n)|/\|g_{n+1}\|^2$ and the eigenvalue by its absolute value. We thus obtain the same results as Theorem 22.1: namely, the inequality (22.7) holds for α_n and β_n, $\{\beta_n\}$ is monotone decreasing and both $\{\alpha_n\}$ and $\{\beta_n\}$ tend to the absolute value of an eigenvalue λ of K, for which $|\lambda| \leq \alpha_n$.

In practice, we usually take β_0 as an approximation to the absolute value of an eigenvalue. This is known as *Rayleigh's principle*. Of course, α_0 is better, as an approximation, than β_0, and α_n or β_n is still better. Furthermore, if we calculate both α and β, then we can estimate the errors of α by virtue of the *Kryloff-Weinstein theorem*, which will be stated later.

REMARK 2. *Remark on Boundary Value Problem.* To calculate α_n and β_n, we need to calculate $g_0(x)$, $g_1(x)$, $g_2(x)$, $\cdots$, $g_{n+1}(x)$ successively. Here $g_m(x)$ is given by

$$g_m(x) = \int_a^b G(x, \xi) g_{m-1}(\xi) d\xi$$

$G(x, \xi)$ denoting the Green's function (or the generalized Green's function). According to Theorem 17.1, we can also determine $g_m(x)$, without constructing the Green's function, by the equation

$$(22.15) \qquad L_x(g_m(x)) = - g_{m-1}(x)$$

together with the boundary conditions in question. For the case when $G(x, \xi)$ is the generalized Green's function, namely, when there exists a non-trivial solution $y_0(x) \not\equiv 0$ of the equation

$$(22.16) \qquad L_x(y_0(x)) = 0$$

satisfying the boundary conditions, we can also determine $g_m(x)$, starting with $g_0(x) \not\equiv 0$ satisfying

$$(22.17) \qquad \int_a^b g_0(x) y_0(x) dx = 0 \,,$$

by the equation (22.15) and the boundary conditions, and using successively the additional condition

$$(22.18) \qquad \int_a^b g_m(x) y_0(x) dx = 0$$

EXAMPLE 1. $y''(x) = -\lambda y(x)$, $y(0) = y(1) = 0$

Take $g_0(x) \equiv 1$. Then, solving the boundary value problem, $g_1''(x) = -1$, $g_1(0) = g_1(1) = 0$, we obtain $g_1(x) = (1/2)(x - x^2)$. From the equation $g_2''(x) = (1/2)(x^2 - x)$, and $g_2(0) = g_2(1) = 0$, we obtain $g_2(x) = (1/24)(x^4 - 2x^3 + x)$. Accordingly, we have

$$\beta_0 = \| g_0 \| / \| g_1 \| = 10.9544$$
$$\alpha_0 = (g_1, g_0) / \| g_1 \|^2 = 10$$
$$\beta_1 = \| g_1 \| / \| g_2 \| = 9.881$$
$$\alpha_1 = (g_2, g_1) / \| g_2 \|^2 = 9.874$$

The approximations α_0, β_0, α_1 and β_1 approach the true eigenvalue

$$\lambda_1 = \pi^2 = 9.8696$$

from above.

EXAMPLE 2. $y''(x) = -\lambda y(x), y(0) = y(1), y'(0) = y'(1)$

In this case there exists a non-trivial solution $y_0(x) \equiv 1$, which satisfies both the differential equation and the boundary conditions. Hence, we take $g_0(x) = 1 - 2x$ which is orthogonal to $y_0(x)$. Then, from $g_1''(x) = 2x - 1$ and $g_1(0) = g_1(1), g_1'(0) = g_1'(1)$, we obtain $g_1(x) = -(1/2)x^2 + (1/3)x^3 + (1/6)x + c$. Since the additional condition $\int_0^1 g_1(x)dx = 0$ implies $c = 0$, we have $g_1(x) = -(1/2)x^2 + (1/3)x^3 + (1/6)x$. Accordingly

$$\beta_0 = \|g_0\| / \|g_1\| = 50.22$$
$$\alpha_0 = (g_1, g_0) / \|g_1\|^2 = 41.99$$

These values α_0 and β_0 are greater than the eigenvalue

$$\lambda_1 = (2\pi)^2 = 39.4784$$

The calculation of α_1 and β_1 is left for the reader.

Kryloff-Weinstein's estimate of the errors.

THEOREM 22.2. Suppose that every eigenvalue of the operator K is positive (or negative). Let α_n and β_n be as in Theorem 22.1. Then, for each n, there exists an eigenvalue λ_{j_n} of K such that

(22.19) $$\sqrt{(\beta_n^2 - \alpha_n^2)} \geqq \alpha_n - \lambda_{j_n} \geqq 0$$

(or

(22.19') $$\sqrt{(\beta_n^2 - \alpha_n^2)} \geqq |\alpha_n| - |\lambda_{j_n}| \geqq 0)$$

and, for sufficiently large n, $\lambda_{j_n} = \lambda_{j_0}$.

Proof. Since $\|g_n - \alpha_n g_{n+1}\|^2 = \|g_n\|^2 - 2\alpha_n(g_n, g_{n+1}) + \alpha_n^2 \|g_{n+1}\|^2$,

$$\|g_n - \alpha_n g_{n+1}\|^2 = \|g_{n+1}\|^2 (\beta_n^2 - \alpha_n^2)$$

Hence, by (22.8) and the Parseval equality (22.3), we obtain

$$\beta_n^2 - \alpha_n^2 = \frac{1}{\|g_{n+1}\|^2} \| \sum_{j=j_0}^{\infty} (\lambda_j^{-(n-1)} - \alpha_n \lambda_j^{-n})(g_1, \varphi_j)\varphi_j \|^2$$

$$= \frac{1}{\|g_{n+1}\|^2} \sum_{j=j_0}^{\infty} (\lambda_j^{-(n-1)} - \alpha_n \lambda_j^{-n})^2 |(g_1, \varphi_j)|^2$$

Accordingly, setting

$$\min_{j \geq j_0}(\lambda_j - \alpha_n)^2 = (\lambda_{j_n} - \alpha_n)^2$$

we obtain, by (22.10), that

$$\beta_n^2 - \alpha_n^2 \geq (\lambda_{j_n} - \alpha_n)^2$$

Since $\lim_{n \to \infty} \beta_n = \lim_{n \to \infty} \alpha_n = \lambda_{j_0}$, we must have $\lambda_{j_n} = \lambda_{j_0}$ for sufficiently large n, q.e.d.

EXAMPLE. (i) The estimate of errors in Example 1 is given by

$$0 < \alpha_0 - \pi^2 = 0.1304 \leqq 4.4721 = (\beta_0^2 - \alpha_0^2)^{1/2}$$

$$0 < \alpha_1 - \pi^2 = 0.0044 \leqq 0.3719 = (\beta_1^2 - \alpha_1^2)^{1/2}$$

(ii) The error estimate in Example 2 is given by

$$0 < \alpha_0 - 4\pi^2 = 2.5116 \leqq 27.539 = (\beta_0^2 - \alpha_0^2)^{1/2}$$

23. *Inhomogeneous integral equations*

We consider the inhomogeneous differential equation

(23.1) $$\frac{d}{dx}\left(p(x)\frac{dy}{dx}\right) - q(x)y = -\lambda r(x)y - s(x), \qquad (r(x) > 0)$$

where λ is a complex parameter and $s(x)$ is a given continuous function. As was shown in the preceding paragraph, the equation (23.1) can be reduced to the integral equation

$$y(x) = \lambda \int_a^b G(x, \xi) r(\xi) y(\xi) d\xi + \int_a^b G(x, \xi) s(\xi) d\xi$$

by means of Green's function $G(x, \xi)$. Since the second term on the right side is continuous, the solution of this integral equation can be reduced to that of the inhomogeneous integral equation

(23.2) $$f(x) = \varphi(x) - \lambda \int_a^b K(x, \xi)\varphi(\xi) d\xi$$

with a real-valued continuous symmetric kernel $K(x, \xi)$ (by setting $K(x, \xi) = \sqrt{(r(x))} G(x, \xi) \sqrt{(r(\xi))}$, and $\varphi(x) = \sqrt{(r(x))} \cdot y(x)$).

In the following, we shall be concerned with the problem of finding the solutions $\varphi(x)$ of the equation (23.2) for a given continuous function $f(x)$ and a given value of complex parameter λ.

As before, we denote by $\{\lambda_j\}$ the eigenvalues of the associated homogeneous equation

(23.3) $$0 = \varphi(x) - \lambda \int_a^b K(x, \xi)\varphi(\xi) d\xi$$

and by $\{\varphi_j(x)\}$ the corresponding complete orthonormal system of eigenfunctions. We have then

(23.4) $$|\lambda_1| \leqq |\lambda_2| \leqq |\lambda_3| \leqq \cdots, \lim_{j \to \infty} |\lambda_j| = \infty$$

and

(23.5) $$K\varphi_j = \lambda_j^{-1}\varphi_j, \quad (\varphi_j, \varphi_k) = \delta_{jk}$$

We shall consider two separate case of the problem. For the first case the value λ is different from any one of the eigenvalues λ_j. For the second case λ is equal to some eigenvalue λ_{j_0}.

First case: λ is different from any one of the eigenvalues λ_j. Suppose that a solution $\varphi(x)$ of (23.2) exists. Then

$$\varphi(x) - f(x) = \lambda \int_a^b K(x, \xi)\varphi(\xi)d\xi$$

Hence, by the expansion theorem, $\varphi(x) - f(x)$ can be expanded in a Fourier series, which converges absolutely and uniformly on the interval $[a, b]$, with respect to the orthonormal system $\{\varphi_j(x)\}$. Let $C_j, j = 1, 2, \cdots$, be the Fourier coefficients of $\varphi(x) - f(x)$. Then

$$\begin{aligned}
C_j &= (\varphi - f, \varphi_j) = (\lambda K\varphi, \varphi_j) = \lambda(K\varphi, \varphi_j) \\
&= \lambda(\varphi, K\varphi_j) = \lambda(\varphi, \lambda_j^{-1}\varphi_j) = \lambda\lambda_j^{-1}(\varphi, \varphi_j) \\
&= \lambda\lambda_j^{-1}(\varphi - f + f, \varphi_j) = \lambda\lambda_j^{-1}(\varphi - f, \varphi_j) + \lambda\lambda_j^{-1}(f, \varphi_j) \\
&= \lambda\lambda_j^{-1}C_j + \lambda\lambda_j^{-1}f_j
\end{aligned}$$

where

$$f_j = (f, \varphi_j)$$

Hence we obtain

(23.6) $$C_j = \lambda f_j/(\lambda_j - \lambda)$$

Consequently, we obtain that a solution $\varphi(x)$, if any, can be expanded as follows:

(23.7) $$\varphi(x) = f(x) + \sum_{j=1}^{\infty} \frac{\lambda}{\lambda_j - \lambda}f_j\varphi_j(x)$$

Actually, the series (23.7) is really a solution of (23.2), that is, it converges absolutely and uniformly on the interval $[a, b]$ and satisfies (23.2). In fact, we obtain, by Bessel's inequality (21.6), that

$$\sum_{j=1}^{\infty}|f_j|^2 \leq \int_a^b |f(x)|^2 dx < \infty$$

$$\sum_{j=1}^{\infty}\lambda_j^{-2}|\varphi_j(x)|^2 \leq \int_a^b |K(x, \xi)|^2 d\xi$$

$$\leq \sup_{a \leq x \leq b}\int_a^b |K(x, \xi)|^2 d\xi < \infty$$

Hence we have

$$\left(\sum_{j=k}^{\infty} \left| \frac{\lambda}{\lambda_j - \lambda} f_j \varphi_j(x) \right| \right)^2 \leq \sum_{j=k}^{\infty} |f_j|^2 \sum_{j=k}^{\infty} \left| \frac{\lambda}{\lambda_j - \lambda} \right|^2 | \varphi_j(x) |^2$$

$$\leq C \sum_{j=k}^{\infty} |f_j|^2 \sum_{j=k}^{\infty} | \lambda_j^{-1} \varphi_j(x) |^2$$

$$\leq C \sum_{j=k}^{\infty} |f_j|^2 \sup_{a \leq x \leq b} \int_a^b | K(x, \xi) |^2 d\xi$$

$$C = \sup_{j \geq 1} \left| \frac{\lambda_j \lambda}{\lambda_j - \lambda} \right|$$

On the other hand, the assumption $\lambda \neq \lambda_j (j = 1, 2, \cdots)$, together with $\lim_{j \to \infty} | \lambda_j | = \infty$, implies $C < \infty$. Thus, we obtain that the series (23.7) converges absolutely and uniformly on the interval.

To prove that the series satisfies the equation (23.2), substitute it on the right side of (23.2) and integrate term by term. Then we obtain that the right side is equal to

$$f(x) + \sum_{j=1}^{\infty} \frac{\lambda}{\lambda_j - \lambda} f_j \varphi_j(x) - \left\{ \lambda \int_a^b K(x, \xi) f(\xi) d\xi \right.$$

$$\left. + \sum_{j=1}^{\infty} \frac{\lambda}{\lambda_j - \lambda} f_j \lambda \int_a^b K(x, \xi) \varphi_j(\xi) d\xi \right\}$$

On the other hand, by the expansion theorem, we have

$$\int_a^b K(x, \xi) f(\xi) d\xi = \sum_{j=1}^{\infty} (Kf, \varphi_j) \varphi_j(x)$$

$$= \sum_{j=1}^{\infty} (f, K\varphi_j) \varphi_j(x) = \sum_{j=1}^{\infty} (f, \lambda_j^{-1} \varphi_j) \varphi_j(x)$$

$$= \sum_{j=1}^{\infty} \lambda_j^{-1} f_j \varphi_j(x)$$

Hence, making use of the identity

$$\frac{\lambda}{\lambda_j - \lambda} - \frac{\lambda}{\lambda_j} - \frac{\lambda^2}{(\lambda_j - \lambda)\lambda_j} = 0,$$

we obtain

$$f(x) + \sum_{j=1}^{\infty} \frac{\lambda}{\lambda_j - \lambda} f_j \varphi_j(x) - \left\{ \sum_{j=1}^{\infty} \lambda_j^{-1} \lambda f_j \varphi_j(x) \right.$$

$$\left. + \sum_{j=1}^{\infty} \frac{\lambda}{\lambda_j - \lambda} f_j \frac{\lambda}{\lambda_j} \varphi_j(x) \right\} = f(x)$$

Accordingly we have the following theorem.

THEOREM 23.1. If λ is different from any one of the eigenvalues of the operator K, and if $f(x)$ is continuous, then the equation (23.2) admits one and only one solution $\varphi(x)$ which is given by (23.7).

Second case: $\lambda = \lambda_{j_0}$ for some j_0. Suppose that $\varphi(x)$ is a solution of the equation

$$(23.2') \qquad f(x) = \varphi(x) - \lambda_{j_0} \int_a^b K(x, \xi)\varphi(\xi)d\xi$$

Multiplying both sides of (23.2') by $\overline{\varphi_j(x)}$ and then integrating from a to b, we obtain

$$f_j = (f, \varphi_j) = (\varphi, \varphi_j) - \lambda_{j_0}(K\varphi, \varphi_j)$$
$$= (\varphi, \varphi_j) - \lambda_{j_0}(\varphi, K\varphi_j) = (\varphi, \varphi_j) - \lambda_{j_0}\lambda_j^{-1}(\varphi, \varphi_j)$$

Hence we see that

$$(23.8) \qquad \text{if } \lambda_j = \lambda_{j_0}, \text{ then } f_j = 0$$

Conversely, if $f(x)$ satisfies (23.8), then we can prove in the same way as in the first case that

$$(23.9) \qquad \varphi(x) = f(x) + \sum_{\lambda_j \neq \lambda_{j_0}} \frac{\lambda}{\lambda_j - \lambda} f_j \varphi_j(x)$$

converges uniformly on the interval $[a, b]$ and satisfies the equation (23.2'); hence $\varphi(x)$ is a particular solution of (23.2'). Furthermore, the difference $\psi(x)$ of this solution $\varphi(x)$ and any other solution, if any, is a solution of the associated homogeneous equation

$$(23.3') \qquad 0 = \psi(x) - \lambda_{j_0} \int_a^b K(x, \xi)\psi(\xi)d\xi$$

hence it can be written as a linear combination

$$\sum_{\lambda_j = \lambda_{j_0}} C_j \varphi_j(x)$$

of the eigenfunctions $\varphi_j(x)$ corresponding to the eigenvalue $\lambda_j = \lambda_{j_0}$. This proves the following

THEOREM 23.2. If $\lambda = \lambda_{j_0}$ for some j_0, then (23.8) is a necessary and sufficient condition for the existence of the solution of the equation (23.2'). If $f(x)$ satisfies (23.8), then the general solution of (23.2') is given by

$$(23.10) \qquad \varphi(x) = \sum_{\lambda_j \neq \lambda_{j_0}} \frac{\lambda f_j}{\lambda_j - \lambda} \varphi_j(x) + \sum_{\lambda_j = \lambda_{j_0}} C_j \varphi_j(x)$$

where the C_j are arbitrary numbers.

REMARK. It should be noted that these results are parallel to the well-known results concerning the solvability of systems of algebraic equations of first order of the form

$$\Psi_t - \lambda \sum_{s=1}^{n} K_{ts}\Psi_s = F_t \quad (t = 1, 2, \cdots, n)$$

$$K_{ts} = K_{st} \qquad\qquad (s, t = 1, 2, \cdots, n)$$

We shall prove, in Chapter 3, Fredholm's theorem which is a generalization of the above theorems to the case of arbitrary kernel $K(x, \xi)$.

24. *Hermite, Laguerre and Legendre polynomials*

In the preceding paragraphs, we were concerned with the eigenvalue problem for the differential operator with real coefficients

$$L_x(y) = \frac{d}{dx}\left(p(x)\frac{dy}{dx}\right) - q(x)y$$

on a closed interval $[a, b]$ under the following assumptions: the interval is a finite one, the coefficients $p(x)$, $q(x)$ are continuous, and $p(x) > 0$. If any one of these assumptions is omitted, the proof of the Hilbert-Schmidt theory can not be carried out as before. For example, when $a = -\infty$, or $b = \infty$, or $a = -\infty$ together with $b = \infty$, it is, in the first place, uncertain what are the suitable boundary conditions for these cases, and whether with such boundary conditions, the boundary value problems can be reduced to integral equations, as before, or not. We are faced with similar difficulties, when either $p(x)$ or $q(x)$ is no longer continuous, or when $p(x)$ is not positive. We will not enter into the details of these problems, which will be called *singular boundary value problems* hereafter and which will be discussed in Chapter 5 from a general point of view. In the following, we shall be chiefly concerned with the Hermite, Laguerre, and Legendre polynomials, which play, in connection with the Schrödinger equation, important roles in quantum mechanics.

Hermite polynomials. We consider the following equation, containing a complex parameter λ,

(24.1) $$(e^{-x^2}y')' + \lambda e^{-x^2}y = 0$$

on the infinite open interval $(-\infty, \infty)$. We take as boundary conditions the following: as $x \to -\infty$, and as $x \to +\infty$, $y(x)$ tends to infinity of an order not greater than a certain finite power of x, that is,

(24.2) $$y(x) = O(x^k) \quad \text{as } x \to \pm\infty$$

The equation (24.1) is obviously equivalent to the equation

$$(24.3) \qquad y'' - 2xy' + \lambda y = 0$$

This equation has no singular points except $x = \pm\infty$, hence its solution $y(x)$ can be given by a power series

$$(24.4) \qquad y(x) = \sum_{n=0}^{\infty} a_n x^n$$

which converges for $|x| < \infty$. Moreover, $y(-x)$ is, with $y(x)$, a solution of (24.3): hence the even function $y(x) + y(-x)$ and the odd function $y(x) - y(-x)$ satisfy the equation (24.3) as well as $y(x)$. Accordingly, it is sufficient to consider only an even and an odd solution

$$(24.4') \qquad y(x) = \sum_{n=0}^{\infty} a_{2n} x^{2n} \quad \text{and} \quad y(x) = \sum_{n=0}^{\infty} a_{2n+1} x^{2n+1}$$

instead of (24.4). Substituting (24.4') in (24.3) and comparing the coefficients of x^n, we obtain the recursion formula

$$(24.5) \qquad a_{n+2}/a_n = (2n - \lambda)/(n + 1)(n + 2)$$

For $y(x)$ to satisfy the boundary conditions (24.2), we must have

$$(24.6) \qquad \lambda = 2n \qquad (n = 0, 1, 2, \cdots)$$

In fact, if it were otherwise, then, by virtue of the recursion formula (24.5), $y(x) = \sum_{n=0}^{\infty} a_{2n} x^{2n}$ (or $\sum_{n=0}^{\infty} a_{2n+1} x^{2n+1}$) is really an infinite series, provided that a_0 (or a_1) is not zero, i.e., if $y(x)$ is a non-trivial solution. Moreover, as will be proved shortly later, every eigenvalue λ for our boundary value problem is real. Hence we have $a_{n+2}/a_n = $ real ($n = 0, 1, 2, \cdots$), because of (24.5). Thus for every n with $2n - \lambda > 0$, a_{2n}/a_0 (or a_{2n+1}/a_1) has the same sign. Therefore such an infinite series $y(x)$ by no means satisfies the boundary conditions (24.2).

Accordingly, we see that all eigenvalues λ are given by (24.6) and the eigenfunction corresponding to the eigenvalue $2n$ is a polynomial of degree n. We define, multiplying these polynomials by certain constants, the nth *Hermite polynomials* $H_n(x)$ as follows:

$$H_0(x) = 1, \quad H_1(x) = 2x, \quad H_2(x) = 4x^2 - 2,$$
$$H_3(x) = 8x^3 - 12, \quad H_4(x) = 16x^4 - 48x^2 + 12,$$

$$(24.7) \qquad \cdots\cdots\cdots\cdots\cdots\cdots\cdots$$

$$H_n(x) = (2x)^n - \frac{n(n-1)}{1!}(2x)^{n-2}$$
$$+ \frac{n(n-1)(n-2)(n-3)}{2!}(2x)^{n-4} - \cdots$$

where the last term is

$$+ (-1)^{\frac{n}{2}} \frac{n!}{(n/2)!} \qquad \text{for even } n$$

and

$$+ (-1)^{\frac{n-1}{2}} \frac{n!}{((n-1)/2)!} 2x \qquad \text{for odd } n.$$

Orthogonality of the Hermite polynomals. The equation (24.1) implies that

$$-2n \int_{-\infty}^{\infty} e^{-x^2} H_n(x) H_m(x) \, dx = \int_{-\infty}^{\infty} (e^{-x^2} H_n'(x))' H_m(x) \, dx$$

By partial integration, we see that the right side is equal to

$$[e^{-x^2} H_n'(x) H_m(x)]_{-\infty}^{\infty} - \int_{-\infty}^{\infty} e^{-x^2} H_n'(x) H_m'(x) \, dx$$

$$= - \int_{-\infty}^{\infty} e^{-x^2} H_n'(x) H_m'(x) \, dx$$

which is symmetric with respect to m and n. Accordingly we have

$$-2n \int_{-\infty}^{\infty} e^{-x^2} H_n(x) H_m(x) \, dx = -2m \int_{-\infty}^{\infty} e^{-x^2} H_n(x) H_m(x) \, dx$$

and therewith

$$(24.8) \qquad \int_{-\infty}^{\infty} e^{-x^2} H_n(x) H_m(x) \, dx = 0 \qquad (m \neq n)$$

showing that the Hermite polynomials form an orthogonal system with the *weight function* e^{-x^2}.

Generating function of the Hermite polynomials. The Hermite polynomials may also be defined by means of a *generating function* $\Psi(x, t)$ as follows

$$(24.9) \qquad \Psi(x, t) = e^{-t^2 + 2tx} = e^{x^2} e^{-(t-x)^2}$$

$$= \sum_{n=0}^{\infty} \frac{H_n(x)}{n!} t^n$$

If we put

$$e^{-t^2 + 2tx} = \sum_{n=0}^{\infty} \frac{C_n(x)}{n!} t^n$$

then we can prove that

$$(24.10) \qquad C_n(x) = (-1)^n e^{x^2} \frac{d^n e^{-x^2}}{dx^n}$$

equals the nth Hermite polynomial. To carry out the proof, differentiate (24.9) with respect to x. We obtain $\Psi_x(x, t) = 2t\Psi(x, t)$, and hence

$$(24.11) \qquad C_n'(x) = 2n\, C_{n-1}(x)$$

On the other hand, differentiating (24.9) with respect to t, we obtain $\Psi_t(x, t) + 2(t - x)\Psi(x, t) = 0$, and hence

$$(24.12) \qquad C_{n+1}(x) - 2x\, C_n(x) + 2n\, C_{n-1}(x) = 0$$

Combining (24.11) and (24.12), we have

$$(24.13) \qquad C_n''(x) - 2x\, C_n'(x) + 2n\, C_n(x) = 0$$

This means that $C_n(x)$ is an eigenfunction corresponding to the eigenvalue $\lambda = 2n$. Since the highest term of the polynomial $C_n(x)$ of degree n is same as that of $H_n(x)$, we must have $C_n(x) \equiv H_n(x)$.

From (24.1) and (24.11), we obtain by partial integration that

$$2n \int_{-\infty}^{\infty} e^{-x^2} H_n^2(x)\, dx = \int_{-\infty}^{\infty} e^{-x^2} (H_n'(x))^2\, dx$$

$$= (2n)^2 \int_{-\infty}^{\infty} e^{-x^2} H_{n-1}^2(x)\, dx$$

Hence we obtain further

$$(24.14) \qquad \int_{-\infty}^{\infty} e^{-x^2} H_n^2(x)\, dx = 2^n(n!) \int_{-\infty}^{\infty} e^{-x^2} H_0^2(x)\, dx$$

$$= 2^n(n!) \int_{-\infty}^{\infty} e^{-x^2}\, dx = 2^n(n!)\sqrt{\pi}$$

Thus the system of the *Hermite functions*

$$\{(2^n(n!)\sqrt{\pi})^{-\frac{1}{2}} e^{-x^2/2} H_n(x)\}$$

is an orthonormal system defined on $(-\infty, \infty)$.

The definition of the completeness of the Hermite polynomials, together with its proof, will be given in Part 51.

REMARK. We shall prove that all eigenvalues of the equation (24.1) under the boundary conditions (24.2) are real.

To prove this, let λ be an eigenvalue and $y(x)$ the corresponding solution. Then, since

$$L_x(y) = (e^{-x^2} y')' = -\lambda e^{-x^2} y$$

the complex conjugate $\overline{y(x)}$ of $y(x)$ satisfies the equation

$$L_x(\bar{y}) = (e^{-x^2}\bar{y}')' = -\bar{\lambda}\,e^{-x^2}\bar{y}$$

and the boundary conditions (24.2). Accordingly, taking account of the boundary conditions, we obtain by partial integration that

$$(\bar{\lambda} - \lambda)\int_{-\infty}^{\infty} e^{-x^2}y(x)\,\overline{y(x)}\,dx = \int_{-\infty}^{\infty} \{\overline{y(x)}\,L_x(y) - y(x)\,L_x(\bar{y})\}\,dx$$

$$= \int_{-\infty}^{\infty} \{\overline{y(x)}\,(e^{-x^2}y')' - y(x)\,(e^{-x^2}\bar{y}')'\}\,dx$$

$$= [\overline{y(x)}\,e^{-x^2}y'(x) - y(x)\,e^{-x^2}\bar{y}'(x)]_{-\infty}^{\infty} = 0$$

Since $y(x) \not\equiv 0$, we thus obtain that $\lambda = \bar{\lambda}$; that is, λ must be real, q.e.d.

Laguerre polynomials. We consider the following differential equation, containing a complex parameter λ,

(24.15) $$(x\,e^{-x}\,y')' + \lambda\,e^{-x}\,y = 0$$

on the infinite interval $[0, \infty)$. We take as boundary conditions the following:

(24.16) $$y(x) \quad \text{remains finite as } x \to 0\,,$$
            $$y(x) = O(x^k) \qquad \text{as } x \to \infty$$

The equation (24.15) is equivalent to the equation

(24.17) $$xy'' + (1 - x)\,y' + \lambda\,y = 0$$

The equation (24.17) has no finite singular points except the point $x = 0$ which is regular singular. The indicial equation at the point $x = 0$ is

(24.18) $$r^2 = 0$$

and has a double root $r = 0$. Hence, there is a solution $y_1(x)$ of (24.17) of the form

(24.19) $$y_1(x) = \sum_{n=0}^{\infty} a_n x^n$$

Substituting this in (24.17) and comparing the coefficient of x^n, we obtain

(24.20) $$a_n = \frac{a_0(-\lambda)(-\lambda + 1)\cdots(-\lambda + n - 1)}{(n!)^2}$$

We shall now seek another solution $y_2(x)$ which is linearly independent of $y_1(x)$ and which is of the form

(24.21) $$y_2(x) = y_1(x) \int^x z(t)\, dt$$

As was shown in (12.29), $z(x)$ satisfies the equation

(24.22) $$z'(x) + \left\{ \frac{1-x}{x} + 2\frac{y_1'(x)}{y_1(x)} \right\} z(x) = 0$$

Hence we obtain that

(24.23)
$$z(x) = \frac{1}{x} e^x y_1(x)^{-2}$$
$$y_2(x) = y_1(x) \int^x \frac{1}{t} e^t y_1(t)^{-2}\, dt$$

Consequently, if a_0 is non-zero and real, that is, $y_1(x)$ is a non-trivial real solution, and if λ is a real number such that

(24.24) $$\lambda \neq n\,, \qquad (n = 0, 1, 2, \cdots)$$

then, on account of (24.20), the coefficients a_n of the infinite series (24.19) have the same sign from a sufficiently large index on; hence the boundary conditions (24.16) are not satisfied. (In view of (24.20), $y_1(x) \neq O(x^k)$ as $x \to \infty$, and moreover $y_2(x)$ does not remain finite as $x \to 0$.) Therefore we see that all the real eigenvalues for our boundary value problem are given by

(24.24′) $$\lambda = 0, 1, 2, \cdots$$

and the solution corresponding to $\lambda = n$ is a polynomial of degree n. The *Laguerre polynomials* $L_n(x)$ are defined by multiplying these polynomials by certain constants as follows:

$$L_0(x) = 1\,, \qquad L_1(x) = -x + 1\,,$$

$$L_2(x) = \frac{x^2}{2} - 2x + 1\,,$$

(24.25) $$L_3(x) = -\frac{x^3}{6} + \frac{3}{2}x^2 - 3x + 1\,,$$

$$L_4(x) = \frac{x^4}{24} - \frac{2}{3}x^3 + 3x^2 - 4x + 1\,,$$

$$\cdots\cdots\cdots\cdots\cdots\cdots$$

$$L_n(x) = \sum_{\nu=0}^{n} \binom{n}{n-\nu} \frac{(-x)^\nu}{\nu!} = \frac{1}{n!} e^x \frac{d^n}{dx^n}(x^n e^{-x})$$

Since $y_2(x)$ does not satisfy the boundary condition at the point $x = 0$, as is easily seen from (24.23), the system of eigenfunctions of our problem is given by the Laguerre polynomials.

The orthogonality of the Laguerre polynomials. We can easily prove the orthogonality relations

$$(24.26) \qquad \int_0^\infty e^{-x} L_n(x) L_m(x) \, dx = 0 \qquad (n \neq m)$$

in the same way as in the case of the Hermite polynomials. Since $L_n(x)$ is a polynomial of degree m, x^m is written as a linear combination of $L_0(x), L_1(x), \cdots, L_m(x)$. Hence (24.26) yields

$$\int_0^\infty e^{-x} x^m L_n(x) \, dx = 0 \qquad (n > m)$$

Accordingly, making use of the fact that

$$L_n(x) = (-1)^n \frac{1}{n!} x^n + \cdots$$

we obtain by repeated partial integration that

$$
\begin{aligned}
\int_0^\infty e^{-x} L_n(x)^2 \, dx &= \int_0^\infty e^{-x} (-1)^n \frac{x^n}{n!} L_n(x) \, dx \\
&= \int_0^\infty (-1)^n \left(\frac{1}{n!}\right)^2 x^n \frac{d^n}{dx^n} (x^n e^{-x}) \, dx \\
&= (-1)^n \left(\frac{1}{n!}\right)^2 (-n) \int_0^\infty x^{n-1} \frac{d^{n-1}}{dx^{n-1}} (x^n e^{-x}) \, dx \\
&= (-1)^{n+2} \left(\frac{1}{n!}\right)^2 n(n-1) \int_0^\infty x^{n-2} \frac{d^{n-2}}{dx^{n-2}} (x^n e^{-x}) \, dx \\
&= \cdots = \frac{1}{n!} \int_0^\infty x^n e^{-x} \, dx = 1
\end{aligned}
$$

We thus obtain that

$$(24.27) \qquad \int_0^\infty e^{-x} L_n(x)^2 \, dx = 1 \qquad (n = 0, 1, 2, \cdots)$$

In other words, the *system of Laguerre functions* $\{e^{-x/2} L_n(x)\}$ is an orthonormal system defined on $[0, \infty)$.

Generating function of Laguerre polynomials. For $|t| < 1$, we have

$$\sum_{n=0}^{\infty} L_n(x)\, t^n = \sum_{n=0}^{\infty} \sum_{\nu=0}^{n} (-1)^\nu \binom{n}{\nu} \frac{1}{\nu!}\, x^\nu t^n$$

$$= \sum_{\nu=0}^{\infty} \frac{(-1)^\nu x^\nu}{\nu!} \sum_{n=\nu}^{\infty} \binom{n}{\nu} t^n$$

$$= \sum_{\nu=0}^{\infty} \frac{(-1)^\nu x^\nu}{\nu!} \frac{t^\nu}{(1-t)^{\nu+1}}$$

$$= \frac{1}{(1-t)} e^{-xt/(1-t)}$$

Hence the Laguerre polynomials may also be defined by means of the *generating function* $1/(1-t)\cdot e^{-xt/(1-t)}$ as follows,

(24.28) $$\sum_{n=0}^{\infty} L_n(x)\, t^n = \frac{1}{1-t} e^{-xt/(1-t)}, \quad |t| < 1$$

We can prove, as in the case of the Hermite polynomials, that all eigenvalues for our problem are real. The definition of the completeness of the Laguerre polynomials, together with its proof, will be given in Part 54.

Legendre polynomials. We consider the following differential equation, containing a complex parameter λ,

(24.29) $$\{(1 - x^2)\, y'\}' + \lambda y = 0$$

on the closed interval $[-1, 1]$. We take as boundary conditions the following:

(24.30)   both $\lim_{x \to -1} y(x)$ and $\lim_{x \to 1} y(x)$ are finite.

As was shown in Part 14,

(24.31) $$\lambda = n(n + 1) \quad (n = 0, 1, 2, \cdots)$$

is the eigenvalue for our boundary value problem and the corresponding eigenfunction is the nth Legendre polynomial $P_n(x)$. We shall prove here the completeness of these polynomials and also prove that there is no eigenvalue except (24.31) and no eigenfunction which is linearly independent of the Legendre polynomials.

Let $f(x)$ be a real-valued continuous function defined on the interval $[-1, 1]$. Then, as was shown in Part 14, $f(x)$ can be approximated uniformly on the interval $[-1, 1]$ by linear combinations of the orthonormal system $\{\psi_n(x)\}$,

(24.32) $$\psi_n(x) = \sqrt{(n + \tfrac{1}{2})} \cdot P_n(x)$$

as follows

$$(24.33) \qquad \sup_{|x| \leq 1} \left| f(x) - \sum_{n=0}^{m} c_n \psi_n(x) \right| < \varepsilon$$

Now, let α_n be the Fourier coefficient of $f(x)$ with respect to the orthonormal system $\{\psi_n(x)\}$,

$$(24.34) \qquad \alpha_n = \int_{-1}^{1} f(x) \, \psi_n(x) \, dx = (f, \psi_n)$$

Then, for any set of numbers $\beta_0, \beta_1, \beta_2, \cdots, \beta_m$, we have

$$\int_{-1}^{1} \left| f(x) - \sum_{n=1}^{m} \beta_n \, \psi_n(x) \right|^2 dx$$

$$= \int_{-1}^{1} |f(x)|^2 \, dx - \sum_{n=0}^{m} \beta_n \, \bar{\alpha}_n - \sum_{n=0}^{m} \bar{\beta}_n \, \alpha_n + \sum_{n=0}^{m} \beta_n \, \bar{\beta}_n$$

$$= \int_{-1}^{1} |f(x)|^2 \, dx - \sum_{n=0}^{m} |\alpha_n|^2 + \sum_{n=0}^{m} |\alpha_n - \beta_n|^2$$

by making use of the orthonormality of the system (24.32). Hence the above integral attains its minimal value when β_n is equal to the Fourier coefficient α_n. Therefore, from (24.33), we obtain

$$2\varepsilon^2 \geq \int_{-1}^{1} \left| f(x) - \sum_{n=0}^{m} c_n \, \psi_n(x) \right|^2 dx$$

$$\geq \int_{-1}^{1} \left| f(x) - \sum_{n=0}^{m} \alpha_n \, \psi_n(x) \right|^2 dx$$

$$= \int_{-1}^{1} |f(x)|^2 \, dx - \sum_{n=0}^{m} |\alpha_n|^2 \geq 0$$

for any positive number ε. Accordingly

$$(24.35) \qquad \int_{-1}^{1} |f(x)|^2 \, dx = \sum_{n=0}^{\infty} \left| \int_{-1}^{1} f(x) \, \psi_n(x) \, dx \right|^2$$

An orthonormal system $\{\psi_n(x)\}$ is said to be *complete* when the completeness relation (or the Parseval equality) (24.35) holds for any continuous function $f(x)$. The Legendre polynomials are, in this sense, complete.

Next we shall prove that (24.31) exhausts all eigenvalues for our problem. To prove this, suppose that there exists an eigenvalue λ' different from (24.31) with the corresponding eigenfunction $\varphi(x)$. It should be noted first that λ' is real, which is proved similarly as in the case of the Hermite polynomials. Further, $\varphi(x)$ is orthogonal to every $\psi_n(x)$. In fact, by the definition of $P_n(x)$, we have

$$\int_{-1}^{1} \{(1 - x^2) P_n'(x)\}' \, \varphi(x) \, dx = - \int_{-1}^{1} n(n + 1) P_n(x) \, \varphi(x) \, dx$$

Then, integrating the left side by parts, we find that it equals

$$[(1 - x^2) P_n'(x) \, \varphi(x)]_{-1}^{1} - \int_{-1}^{1} (1 - x^2) P_n'(x) \, \varphi'(x) \, dx$$

$$= \int_{-1}^{1} -(1 - x^2) P_n'(x) \, \varphi'(x) \, dx$$

and hence equals

$$\int_{-1}^{1} \{(1 - x^2) \varphi'(x)\}' P_n(x) \, dx = - \int_{-1}^{1} \lambda' \, \varphi(x) P_n(x) \, dx$$

Thus we obtain, for $\lambda' \neq n(n + 1)$ $(n = 0, 1, 2, \cdots)$,

(24.36) $$\int_{-1}^{1} P_n(x) \, \varphi(x) \, dx = \int_{-1}^{1} \psi_n(x) \, \varphi(x) \, dx = 0 \qquad (n = 0, 1, 2, \cdots)$$

Setting $f(x) = \varphi(x)$ in (24.35), we see that the continuous function $\varphi(x)$ satisfies

$$\int_{-1}^{1} |\varphi(x)|^2 \, dx = 0$$

hence $\varphi(x) \equiv 0$: this is a contradiction.

Consequently, we can conclude that all eigenvalues for the boundary value problem (24.29)–(24.30) are given by (24.31), and the corresponding eigenfunctions are the functions $\psi_n(x)$.

REMARK. The completeness relation (24.35) may be stated in the following, apparently more general, form: the system $\{\psi_n\}$ is complete if for any pair of real-valued continuous functions $f(x)$, $g(x)$ defined on the interval $[-1, 1]$, there holds the relation

(24.37) $$\int_{-1}^{1} f(x) \, g(x) \, dx = \sum_{n=0}^{\infty} \left(\int_{-1}^{1} f(x) \, \psi_n(x) \, dx \right) \left(\int_{-1}^{1} g(s) \, \psi_n(s) \, ds \right)$$

Proof. Setting $(f, \psi_n) = f_n$, $(g, \psi_n) = g_n$ and by making use of the Schwarz inequality, we obtain

$$\left| (f, g) - \left(f, \sum_{n=0}^{m} g_n \psi_n \right) \right|^2 = \left| \left(f, g - \sum_{n=0}^{m} g_n \psi_n \right) \right|^2$$

$$\leq \| f \|^2 \left\| g - \sum_{n=0}^{m} g_n \psi_n \right\|^2$$

The second term on the right side tends to zero as $m \to \infty$, on account of (24.35). Hence we obtain

$$(f, g) = \lim_{m \to \infty} \left(f, \sum_{n=0}^{m} g_n \psi_n \right) = \lim_{m \to \infty} \sum_{n=0}^{m} g_n(f, \psi_n) = \sum_{n=0}^{\infty} g_n f_n$$

Concerning the details of "orthogonal polynomials," the reader is referred to G. Szegö, *Orthogonal polynomials*, New York, 1939.

§3.   Asymptotic expression of eigenvalues and eigenfunctions. Liouville's method

25.   *Liouville transformation*

We consider the equation

(25.1) $$\frac{d}{dx}\left(p(x)\frac{dy}{dx} \right) + \{\lambda r(x) - q(x)\} y = 0$$

where $p(x)$, $q(x)$ and $r(x)$ are real-valued continuous functions defined on the closed interval $[a, b]$, and $p(x)$, $r(x)$ are positive at any point x on the interval.   If, in addition, $\{p(x) r(x)\}''$, together with $p'(x)$, is continuous on the interval $[a, b]$, then the equation (25.1) is transformed by the *Liouville transformation*

(25.2)
$$z = \frac{1}{K}\int_{a}^{x}\left(\frac{r(x)}{p(x)} \right)^{\frac{1}{2}} dx$$
$$K = \frac{1}{\pi}\int_{a}^{b}\left(\frac{r(x)}{p(x)} \right)^{\frac{1}{2}} dx$$
$$u = (p(x) r(x))^{\frac{1}{4}} y$$
$$\rho^2 = K^2 \lambda$$

into

(25.3) $$\frac{d^2u}{dz^2} + \{\rho^2 - g(z)\} u = 0$$

where

(25.4)
$$g(z) = \frac{f''(z)}{f(z)} - K^2 k(z)$$
$$f(z) = (p(x) r(x))^{\frac{1}{4}}, \qquad k(z) = q(x)/r(x)$$

and the new independent variable z varies over the interval $0 \leq z \leq \pi$.

We shall be primarily concerned with the equation (25.3) under the boundary conditions

(25.5) $$u'(0) - h u(0) = 0 , \qquad u'(\pi) + H u(\pi) = 0 \qquad (h > 0, H > 0)$$

which correspond to

$$y'(a) - \alpha\,y(a) = 0\,, \qquad y'(b) + \beta\,y(b) = 0 \qquad (\alpha > 0,\ \beta > 0)$$

for the original equation (25.1). We write the equation (25.3) as

(25.6) $$d^2u/dz^2 + \rho^2 u = g(z)u$$

For a given continuous function $v(z)$, the equation

$$u''(z) + \rho^2 u(z) = v(z)$$

has a particular solution

$$u(z) = \frac{1}{\rho}\int_0^z [\sin \rho(z - t)]\, v(t)\, dt$$

Hence the general solution of (25.6) can be written in the form

$$u(z) = A \cos \rho z + B \sin \rho z + \frac{1}{\rho}\int_0^z [\sin \rho(z - t)]\, g(t)u(t)\, dt$$

where A, B are integration constants. Accordingly, the solution satisfying the following initial conditions at the point $z = 0$

(25.7) $$u(0) = 1\,, \qquad u'(0) = h$$

is

(25.8) $$u(z) = \cos \rho z + \frac{h}{\rho}\sin \rho z + \frac{1}{\rho}\int_0^z [\sin \rho(z - t)]\, g(t)\, u(t)\, dt$$

If we impose on $u(z)$ a further condition at the point $z = \pi$

(25.9) $$u'(\pi) + H u(\pi) = 0$$

then ρ is determined by

(25.10) $$\tan \rho\pi = C/(\rho - D)$$

where

(25.11)
$$C = h + H + \int_0^\pi \left\{\cos \rho\, t - \frac{H}{\rho}\sin \rho\, t\right\} g(t)\, u(t)\, dt$$
$$D = H\frac{h}{\rho} + \int_0^\pi \left\{\sin \rho\, t + \frac{H}{\rho}\cos \rho\, t\right\} g(t)\, u(t)\, dt$$

The details of the calculation are left for the reader.

No eigenfunction $\varphi(z)$ corresponding to (25.5) and (25.6) vanishes at $z = 0$, for if otherwise, $\varphi(0) = \varphi'(0) = 0$ which implies $\varphi(z) \equiv 0$. Hence the eigenfunction $u(z) = \varphi(z)/\varphi(0)$ satisfies (25.7), and we may assume that, for an eigenvalue and the corresponding eigenfunction $u(z)$, the relations (25.8), (25.10) and (25.11) hold.[1] We know already, as was shown in the preceding paragraphs, that the eigenvalue, together with the corresponding eigenfunction, does exist and the set of all eigenvalues $\{\rho^2\}$ is denumerable and has no limiting points except $\pm\infty$. But $-\infty$ is excluded since, by partial integration of (25.3),

$$\rho^2 \int_0^\pi u^2 dz = \int_0^\pi gu^2 dz - \int_0^\pi u''u dz = \int_0^\pi gu^2 dz - [u'u]_0^\pi + \int_0^\pi (u')^2 dz$$

and $-[u'u]_0^\pi = Hu(\pi)^2 + hu(0)^2 > 0$ by $h > 0$, $H > 0$.

26. Asymptotic expressions of eigenvalues and eigenfunctions

It should be noted first that $u(z)$ is bounded on the interval $[0, \pi]$, that is, $\sup_{0 \leq z \leq \pi} |u(z)| = M < \infty$, since $u(z)$ is a solution of the equation (25.3) satisfying the initial conditions (25.7). Hence it follows from (25.8) that

$$|u(z)| \leq \left(1 + \frac{h^2}{\rho^2}\right)^{\frac{1}{2}} + \frac{M}{\rho} \int_0^\pi |g(t)|\, dt$$

and therewith

$$M \leq \left(1 + \frac{h^2}{\rho^2}\right)^{\frac{1}{2}} + \frac{M}{\rho} \int_0^\pi |g(t)|\, dt$$

We thus obtain

(26.1)
$$M \leq (1 + h^2/\rho^2)^{\frac{1}{2}} \Big/ \left\{1 - (1/\rho)\int_0^\pi |g(t)|\, dt\right\}$$

whenever $\quad \rho > \int_0^\pi |g(t)|\, dt$.

Accordingly, since $M \leq 1 + O(\rho^{-1})$ as $\rho \to \infty$, we see that both $|C|$ and $|D|$ in (25.11) are bounded for $\rho \to \infty$. Hence, by virtue of (25.10), $\rho \approx n$, where n denotes a positive integer, as $\rho \to \infty$.

In the following, we assume further that $g(t)$ is continuously differentiable, and on the basis of this we shall obtain more precise asymptotic estimates for ρ^2 and $u(z)$ as $\rho \to \infty$.

Substituting $u(z) = 1 + O(\rho^{-1})$ in (25.8), we see immediately that, for sufficiently large ρ, $u(z)$ may be written as

$$u(z) = \cos \rho z + \frac{a(\rho, z)}{\rho}$$

---

[1] The equation (25.8), which contains the unknown $u(z)$ in the indefinite integral, is an example of a Volterra integral equation.

where $|a(\rho, z)|$ is bounded when $\rho \to \infty$ for all $z \in [0, \pi]$ simultaneously
Substituting this in (25.8) once more, we obtain

$$u(z) = \left\{1 - \frac{1}{\rho}\int_0^z \sin \rho\, t\left(\cos \rho\, t + \frac{a(\rho, t)}{\rho}\right)g(t)\, dt\right\}\cos \rho z$$

$$+ \left\{\frac{h}{\rho} + \frac{1}{\rho}\int_0^z \cos \rho\, t\left(\cos \rho\, t + \frac{a(\rho, t)}{\rho}\right)g(t)\, dt\right\}\sin \rho z$$

On the other hand, we obtain by partial integration that

$$\int_0^z \sin \rho\, t\cdot\cos \rho\, t\cdot g(t)\, dt = \left[\frac{\sin^2 \rho\, t}{2\rho}\, g(t)\right]_0^z - \int_0^z \frac{\sin^2 \rho\, t}{2\rho}\, g'(t)\, dt = O(\rho^{-1})$$

$$\int_0^z \cos^2 \rho\, t\cdot g(t)\, dt = \int_0^z \frac{1}{2}\, g(t)\, dt + \int_0^z \left(\cos^2 \rho\, t - \frac{1}{2}\right)g(t)\, dt$$

$$= \frac{1}{2}\int_0^z g(t)\, dt + \left[\frac{\sin \rho\, t\cdot\cos \rho\, t}{2\rho}\, g(t)\right]_0^z - \int_0^z \frac{\sin \rho\, t\cdot\cos \rho\, t}{2\rho}\, g'(t)\, dt$$

Hence we obtain

(26.2)
$$u(z) = [1 + O(\rho^{-2})]\cos \rho z + \left[\frac{G(z)}{\rho} + O(\rho^{-2})\right]\sin \rho z$$

$$G(z) = h + \frac{1}{2}\int_0^z g(t)\, dt$$

We similarly obtain from (25.11)

(26.3)
$$C = h + H + h_1 + O(\rho^{-1})\ , \quad h_1 = \frac{1}{2}\int_0^\pi g(t)\, dt$$

$$D = O(\rho^{-1})$$

Hence, according to (25.10), we have

(26.4)
$$\tan \pi\rho = \frac{h + H + h_1 + O(\rho^{-1})}{\rho + O(\rho^{-1})}$$

Thus we see that for sufficiently large values of ρ

$$\rho = n + \frac{h + H + h_1}{\pi\rho} + O(\rho^{-2})$$

$$= n + \frac{h + H + h_1}{\pi n} + O(n^{-2})$$

n denoting a positive integer. Accordingly, we have

(26.5)
$$\rho_n = n + \frac{c}{n} + O(n^{-2})\ , \quad c = (h + H + h_1)/\pi$$

Thus

$$\cos \rho_n z = [1 + O(n^{-2})] \cos nz - \left[\frac{cz}{n} + O(n^{-2}) \right] \sin nz$$

$$\sin \rho_n z = [1 + O(n^{-2})] \sin nz + \left[\frac{cz}{n} + O(n^{-2}) \right] \cos nz$$

and hence we see, by (26.2), that the eigenfunction $u_n(z)$ corresponding to the eigenvalue ρ_n^2 is given by

$$(26.6) \qquad u_n(z) = [1 + O(n^{-2})] \cos nz + \left[\frac{G(z) - cz}{n} + O(n^{-2}) \right] \sin nz$$

We thus obtain the asymptotic expressions (26.5) and (26.6) corresponding to the boundary conditions (25.5). In a similar way, we can obtain asymptotic expressions of the eigenvalues corresponding to the boundary conditions

$$(26.7) \qquad u(0) = 0 , \qquad u'(\pi) + Hu(\pi) = 0$$

$$(\text{or} \quad u'(0) - h u(0) = 0 , \quad u(\pi) = 0)$$

and

$$(26.8) \qquad u(0) = 0 , \qquad u(\pi) = 0$$

respectively. They are

$$(26.9) \qquad \rho_n = n + \tfrac{1}{2} + O(n^{-1})$$

for (26.7), and

$$(26.10) \qquad \rho_n = n + 1 + O(n^{-1})$$

for (26.8).

The asymptotic expressions of the eigenvalues and eigenfunctions for the equation (25.1) can be derived from the above expressions by making use of (25.2).

REMARK. Part 26 is adapted from E. L. Ince: Ordinary Differential Equations, referred to in the Bibliography.

CHAPTER 3

FREDHOLM INTEGRAL EQUATIONS

Let $K(s, t)$ be a complex-valued continuous function defined on a domain $a \leqq s \leqq b$, $a \leqq t \leqq b$; $f(s)$ a complex-valued continuous function defined on the interval $a \leqq s \leqq b$. Further, let λ be a complex parameter. We consider an equation, in an unknown $\varphi(s)$, of the form

$$f(s) = \varphi(s) - \lambda \int_a^b K(s, t)\varphi(t)\, dt$$

This equation is called a *Fredholm integral equation of the second kind*. In this chapter, we shall be chiefly concerned with the problem to find a continuous solution $\varphi(s)$ of the above equation. We have previously discussed in Chapter 2 the class of equation of this form for which the kernel is symmetric, that is, for which

$$K(s, t) = \overline{K(t, s)}$$

We will also be able to derive, without assuming the symmetry property of the kernel, the results corresponding to Theorems 23.1 and 23.2, which are known as the *Fredholm alternative theorem*. We shall start with proving this theorem by Schmidt's method.

§ 1. Fredholm alternative theorem

27. *The case when* $\displaystyle\int_a^b \int_a^b |K(s, t)|^2\, ds\, dt < 1$

For the sake of simplicity, we take $\lambda = 1$ and consider the equation

(27.1) $$\varphi(s) - \int_a^b K(s, t)\varphi(t)\, dt = f(s)$$

An equation, in the unknown $\psi(t)$, of the form

(27.2) $$\psi(t) - \int_a^b K(s, t)\psi(s)\, ds = g(t)$$

$g(t)$ being a given continuous function on the interval $a \leqq t \leqq b$, is called the *conjugate equation* of (27.1).

THEOREM 27. 1. Under the assumption

$$(27.3) \qquad \int_a^b \int_a^b |K(s, t)|^2 \, ds \, dt < 1$$

the equation (27.1) [(27.2)] admits one and only one solution $\varphi(s)[\psi(t)]$ for any $f(s)[g(t)]$; in particular, $\varphi(s) \equiv 0[\psi(t) \equiv 0]$ for the homogeneous equation

$$(27.4) \qquad \varphi(s) - \int_a^b K(s, t)\varphi(t) \, dt = 0$$

$$(27.5) \qquad \psi(t) - \int_a^b K(s, t)\psi(s) \, ds = 0$$

Proof. Starting with the kernel $K(s, t)$, we define the *iterated kernels* $K^{(1)}(s, t)$, $K^{(2)}(s, t)$, $\cdots$, $K^{(n)}(s, t)$, $\cdots$ as follows:

$$K^{(1)}(s, t) = K(s, t)$$

$$(27.6) \qquad K^{(2)}(s, t) = \int_a^b K(s, r)K(r, t) \, dr$$

$$\cdots \quad \cdots \quad \cdots$$

$$K^{(n)}(s, t) = \int_a^b K(s, r)K^{(n-1)}(r, t) \, dr$$

The following relation clearly holds for the iterated kernels

$$(27.7) \qquad K^{(n+m)}(s, t) = \int_a^b K^{(n)}(s, r)K^{(m)}(r, t) \, dr$$

By (27.6) and the Schwarz inequality, we have

$$|K^{(n)}(s, t)|^2 \leqq \int_a^b |K(s, r)|^2 \, dr \int_a^b |K^{(n-1)}(r, t)|^2 \, dr$$

hence

$$\int_a^b \int_a^b |K^{(n)}(s, t)|^2 \, ds \, dt$$

$$\leqq \int_a^b \int_a^b |K(s, r)|^2 \, ds \, dr \int_a^b \int_a^b |K^{(n-1)}(r, t)|^2 \, dr \, dt$$

Repeating this procedure, we finally obtain

$$(27.8) \qquad \int_a^b \int_a^b |K^{(n)}(s, t)|^2 \, ds \, dt \leqq \left[\int_a^b \int_a^b |K(s, t)|^2 \, ds \, dt \right]^n$$

On the other hand, according to (27.6) and (27.7), we see that, for $n \geqq 3$,

$$K^{(n)}(s, t) = \int_a^b \int_a^b K(s, r)K^{(n-2)}(r, r_1)K(r_1, t) \, dr \, dr_1$$

Hence by the Schwarz inequality we have

$$| K^{(n)}(s, t) |^2$$
$$\leqq \int_a^b \int_a^b | K^{(n-2)}(r, r_1) |^2 \, dr \, dr_1 \int_a^b \int_a^b | K(s, r)K(r_1, t) |^2 \, dr \, dr_1$$

Accordingly, by making use of (27.8), we obtain

$$| K^{(n)}(s, t) |^2$$
$$\leqq \left[\int_a^b \int_a^b | K(s, t) |^2 \, ds \, dt \right]^{n-2} \left\{ \int_a^b | K(s, r) |^2 \, dr \int_a^b | K(r_1, t) |^2 \, dr_1 \right\}$$

The term in braces on the right side is continuous on the domain $a \leqq s \leqq b, a \leqq t \leqq b$; hence bounded. Therefore, according to the assumption (27.3), the series

$$(27.9) \qquad \Gamma(s, t) = \sum_{n=1}^{\infty} K^{(n)}(s, t)$$

converge uniformly on the domain $a \leqq s \leqq b, a \leqq t \leqq b$. Hence by term-by-term integration and by using (27.7) we obtain

$$(27.10) \qquad \Gamma(s, t) = K(s, t) + \int_a^b K(s, r)\Gamma(r, t) \, dr$$

$$(27.11) \qquad \Gamma(s, t) = K(s, t) + \int_a^b \Gamma(s, r)K(r, t) \, dr$$

The series (27.9) is known as the *Neumann series* for the kernel $K(s, t)$.

Now, by making use of (27.10), we can prove that

$$(27.12) \qquad \varphi(s) = f(s) + \int_a^b \Gamma(s, t)f(t) \, dt$$

satisfies the equation (27.1). In fact, substituting (27.12) in (27.1) and using (27.10), we have

$$\varphi(s) - \int_a^b K(s, t)\varphi(t) \, dt$$
$$= f(s) + \int_a^b \Gamma(s, t)f(t) \, dt$$
$$- \int_a^b K(s, t) \left\{ f(t) + \int_a^b \Gamma(t, r)f(r) \, dr \right\} dt$$

$$= f(s) + \int_a^b \left\{ \Gamma(s, t) - K(s, t) - \int_a^b K(s, r) \Gamma(r, t) \, dr \right\} f(t) \, dt$$

$$= f(s)$$

Conversely, we can prove that if $\varphi(s)$ satisfies the equation (27.1), then $\varphi(s)$ satisfies (27.12). In fact, substituting $f(s) = \varphi(s) - \int_a^b K(s, t)\varphi(t) \, dt$ in (27.12) and using (27.11), we see that

$$\varphi(s) - \int_a^b K(s, t)\varphi(t) \, dt$$

$$+ \int_a^b \Gamma(s, t) \left\{ \varphi(t) - \int_a^b K(t, r) \, \varphi(r) \, dr \right\} dt$$

$$= \varphi(s) + \int_a^b \left\{ \Gamma(s, t) - K(s, t) - \int_a^b \Gamma(s, r) K(r, t) \, dr \right\} \varphi(t) \, dt$$

$$= \varphi(s)$$

Accordingly, we see that the equation (27.1) is equivalent to the equation (27.12). Similarly, we can prove that the conjugate equation (27.2) is equivalent to the equation

$$(27.13) \qquad \psi(t) = g(t) + \int_a^b \Gamma(s, t)g(s) \, ds$$

q.e.d.

REMARK. Under the assumption (27.3), every solution $\varphi(s)$ of the equation (27.1) is given by (27.12) by means of the kernel $\Gamma(s, t)$ and every solution $\psi(t)$ of the conjugate equation (27.2) is given by (27.13) by means of the *conjugate kernel* $\Gamma'(s, t)$ of $\Gamma(s, t)$, defined by

$$(27.14) \qquad \Gamma'(s, t) = \Gamma(t, s)$$

For this reason, the kernels $\Gamma(s, t)$ and $\Gamma'(s, t)$ are called the *resolvent kernels* of the equations (27.1) and (27.2) respectively.

The foregoing theorem shows that 1 is not an eigenvalue of either the kernel $K(s, t)$ or its conjugate kernel $K'(s, t)$,

$$(27.15) \qquad K'(s, t) = K(t, s)$$

28. *The general case*

We shall prove that there exist two sets of linearly independent continuous functions

$$(28.1) \qquad \begin{array}{l} \alpha_1(s), \ \alpha_2(s), \ \cdots, \ \alpha_m(s) \\ \beta_1(t), \ \beta_2(t), \ \cdots, \ \beta_m(t) \end{array}$$

defined on the interval $[a, b]$, such that

$$(28.2) \qquad \int_a^b \int_a^b \left| K(s, t) - \sum_{\nu=1}^m \alpha_\nu(s)\beta_\nu(t) \right|^2 ds\, dt < 1$$

To prove this, let ε be an arbitrary positive number. Then we divide the interval (a, b) into a finite number of subintervals $I_1, I_2, \cdots, I_n$, such that

$$\sup_{a \leq s \leq b} | K(s, t') - K(s, t'') | \leq \varepsilon$$

for any pair of points t', t'' in each I_ν. This is possible, because of the uniform continuity of $K(s, t)$ on the domain $a \leq s \leq b$, $a \leq t \leq b$. Let t_ν be an inner point of I_ν. Let I_ν' be an interval contained in the interior of I_ν and containing the point t_ν. Then we define $\beta_\nu(t)$ as follows:

$$\beta_\nu(t) \equiv \begin{cases} 0 & \text{outside of } I_\nu \\ 1 & \text{on } I_\nu' \end{cases}$$

such that the function $\beta_\nu(t)$ is continuous and $0 \leq \beta_\nu(t) \leq 1$ on the interval $[a, b]$. We now set $\alpha_\nu(s) \equiv K(s, t_\nu)$ and

$$N(s, t) = \left| K(s, t) - \sum_{\nu=1}^n \alpha_\nu(s)\beta_\nu(t) \right|$$

Then we see that

$$| N(s, t) | = | K(s, t) - K(s, t_\nu) | \leq \varepsilon$$

for t in I_ν', and

$$| N(s, t) | = | K(s, t) - K(s, t_\nu)\beta_\nu(t) | \leq 2M$$

for t in $I_\nu - I_\nu'$ where

$$M = \sup_{a \leq s \leq b, a \leq t \leq b} | K(s, t) |$$

Since ε and the sum of lengthes of $I_\nu - I_\nu'$ are both arbitrary, we can choose the values of them so small that

$$\int_a^b \int_a^b \left| K(s, t) - \sum_{\nu=1}^n \alpha_\nu(s)\beta_\nu(t) \right|^2 ds\, dt < 1$$

Clearly, the functions $\beta_1(t)$, $\beta_2(t)$, $\cdots$, $\beta_n(t)$ are linearly independent. Hence, if $\alpha_1(s)$, $\alpha_2(s)$, $\cdots$, $\alpha_n(s)$ are linearly independent, then our proof is completed. If otherwise, say, $\alpha_n(s)$ is written as a linear combination of $\alpha_1(s)$, $\alpha_2(s)$, $\cdots$, $\alpha_{n-1}(s)$, then

$$R(s, t) \equiv \sum_{\nu=1}^n \alpha_\nu(s)\beta_\nu(t)$$

is also written in the form

$$R(s, t) \equiv \sum_{\nu=1}^{n-1} \alpha_\nu(s)\beta_\nu^{(1)}(t)$$

If $\beta_1^{(1)}(t)$, $\beta_2^{(1)}(t)$, $\cdots$, $\beta_{n-1}^{(1)}(t)$ are linearly independent, then, by setting $\beta_\nu^{(1)}(t) = \beta_\nu(t)$, the number n is diminished. If otherwise, say, $\beta_{n-1}^{(1)}(t)$ is written as a linear combination of $\beta_1^{(1)}(t)$, $\beta_2^{(2)}(t)$, $\cdots$, $\beta_{n-2}^{(1)}(t)$, then $R(s, t)$ is also written as

$$R(s, t) \equiv \sum_{\nu=1}^{n-2} \alpha_\nu^{(1)}(s)\beta_\nu^{(1)}(t)$$

Repeating this argument alternatively for α and β, we finally obtain two sets of linearly independent functions

$$\gamma_1(s),\ \gamma_2(s),\ \cdots,\ \gamma_m(s),\ \text{and}\ \delta_1(t),\ \delta_2(t),\ \cdots,\ \delta_m(t)$$

in terms of which $R(s, t)$ is written as

$$R(s, t) \equiv \sum_{\nu=1}^{m} \gamma_\nu(s)\delta_\nu(t)$$

provided that $K(s, t) \not\equiv 0$ and $R(s, t) \not\equiv 0$. Then, by setting $\gamma_\nu(s) = \alpha_\nu(s)$, and $\delta_\nu(t) = \beta_\nu(t)$, the proof is completed, q.e.d.

We now set

$$K_1(s, t) = K(s, t) - \sum_{\nu=1}^{m} \alpha_\nu(s)\beta_\nu(t)$$

and denote the resolvent kernel of $K_1(s, t)$ by

$$\Gamma_1(s, t) = \sum_{n=1}^{\infty} K_1^{(n)}(s, t)$$

Then, the equation (27.1) is written as

$$(28.3) \qquad \varphi(s) - \int_a^b K_1(s, t)\varphi(t)\, dt$$

$$= f(s) + \int_a^b \left(\sum_{\nu=1}^{m} \alpha_\nu(s)\beta_\nu(t) \right) \varphi(t)\, dt$$

and we can prove in the same way as in Part 27 that $\varphi(s)$ is determined by

$$(28.4) \qquad \varphi(s) - \int_a^b \left[\sum_{\nu=1}^{m} \left(\alpha_\nu(s) + \int_a^b \Gamma_1(s, r)\, \alpha_\nu(r)\, dr \right)\beta_\nu(t) \right]\varphi(t)\, dt$$

$$= f(s) + \int_a^b \Gamma_1(s, r)f(r)\, dr$$

From this follows the fact that to solve the equation (27.1) is equivalent to finding a solution $\varphi(s)$ of the equation (28.4) with the term in brackets as the kernel and for the right side the given function

$$f(s) + \int_a^b \Gamma_1(s, r)f(r)\,dr$$

We shall prove incidentally that

(28.5) $$\qquad \alpha_\nu(s) + \int_a^b \Gamma_1(s, r)\alpha_\nu(r)\,dr \qquad (\nu = 1, 2, \cdots, m)$$

are linearly independent. To prove this, suppose

$$0 \equiv \sum_{\nu=1}^m c_\nu\left(\alpha_\nu(s) + \int_a^b \Gamma_1(s, r)\,\alpha_\nu(r)\,dr\right)$$

$$= \sum_{\nu=1}^m c_\nu\alpha_\nu(s) + \int_a^b \Gamma_1(s, r)\left(\sum_{\nu=1}^m c_\nu\alpha_\nu(r)\right)dr$$

and $\sum_{\nu=1}^m |c_\nu| \neq 0$. Then, by the properties of the resolvent kernel $\Gamma_1(s, t)$, we have

$$\sum_{\nu=1}^m c_\nu\alpha_\nu(s) \equiv 0 - \int_a^b K_1(s, r) \cdot 0\,dr \equiv 0$$

This contradicts the linear independence of $\alpha_\nu(s)$.

The equation (28.4) is reduced to the system of equations

(28.6) $$\qquad \varphi(s) = f(s) + \int_a^b \Gamma_1(s, r)f(r)\,dr$$

$$+ \sum_{\nu=1}^m \rho_\nu\left(\alpha_\nu(s) + \int_a^b \Gamma_1(s, r)\,\alpha_\nu(r)\,dr\right)$$

(28.7) $$\qquad \rho_\mu = \int_a^b \beta_\mu(t)\varphi(t)\,dt \qquad\qquad (\mu = 1, 2, \cdots, m)$$

Hence, substituting (28.6) in (28.7), we have a system of linear equations in unknowns $\rho_1, \rho_2, \cdots, \rho_m$,

(28.8) $$\quad \rho_\mu - \sum_{\nu=1}^m \rho_\nu\left[\int_a^b \alpha_\nu(s)\beta_\mu(s)\,ds + \int_a^b \Gamma_1(s, r)\,\alpha_\nu(r)\beta_\mu(s)\,dr\,ds\right]$$

$$= \int_a^b \left(\beta_\mu(t) + \int_a^b \Gamma_1(r, t)\,\beta_\mu(r)\,dr\right)f(t)\,dt \qquad (\mu = 1, 2, \cdots, m)$$

Accordingly to solve the equation (27.1) is equivalent to finding the solutions ρ_μ of (28.8); indeed, substituting the solutions ρ_μ in (28.6), we obtain the solution of (27.1).

Similarly, we see that to solve the equation (27.2) is equivalent to solving the following system of linear equations in the unknowns

$\rho_1',\ \rho_2',\ \cdots,\ \rho_m',$

(28.9) $$\rho_\mu' - \sum_{\nu=1}^{m} \rho_\nu' \left[\int_a^b \alpha_\mu(t)\beta_\nu(t)\, dt \right.$$
$$\left. + \int_a^b \Gamma_1(r,\ t)\alpha_\mu(t)\beta_\nu(r)\, dr\, dt \right]$$
$$= \int_a^b \left(\alpha_\mu(s) + \int_a^b \Gamma_1(s,\ r)\, \alpha_\mu(r)\, dr \right) g(s)ds \quad (\mu=1, 2, \cdots, m)$$

and the solution $\psi(t)$ of (27.2) is given by

(28.10) $$\psi(t) = g(t) + \int_a^b \Gamma_1(r,\ t)g(r)\, dr$$
$$+ \sum_{\nu=1}^{m} \rho_\nu' \left(\beta_\nu(t) + \int_a^b \Gamma_1(r,\ t)\, \beta_\nu(r)\, dr \right)$$

where the ρ_ν' are the solutions of (28.9).

Let Δ be the matrix of the equations (28.8), in the unknowns ρ, and Δ' that of the equations (28.9), in the unknowns ρ'. Then

(28.11) Δ' is the transposed matrix of Δ

Hence det $\Delta' \neq 0$ if and only if det $\Delta \neq 0$.

We first consider the case when det $\Delta \neq 0$, and hence, det $\Delta' \neq 0$. In this case, the equation (28.8) [(28.9)] for any function $f(s)[g(t)]$, admits a unique solution

$$\rho = (\rho_1,\ \rho_2,\ \cdots,\ \rho_m)[\rho' = (\rho_1',\ \rho_2',\ \cdots,\ \rho_m')]$$

Therefore, for the given function $f(s)[g(t)]$, the equation (27.1) [(27.2)] admits a unique solution $\varphi(s)[\psi(t)]$. In particular, if $f(s) \equiv 0$ $[g(t)\equiv 0]$, then

$$\rho = (\rho_1,\ \rho_2,\ \cdots,\ \rho_m) = (0,\ 0,\ \cdots,\ 0)$$
$$[\rho' = (\rho_1',\ \rho_2',\ \cdots,\ \rho_m') = (0,\ 0,\ \cdots,\ 0)]$$

hence $\varphi(s) \equiv 0$ $[\psi(t) \equiv 0]$.

We next consider the case when det $\Delta = 0$, and hence det $\Delta' = 0$. For the sake of simplicity, we write (28.8), (28.9) as

(28.8′) $$\rho_\mu - \sum_{\nu=1}^{m} c_{\mu\nu}\rho_\nu = f_\mu \qquad\qquad (\mu = 1, 2, \cdots, m)$$

(28.9′) $$\rho_\mu' - \sum_{\nu=1}^{m} c_{\nu\mu}\rho_\nu' = g_\mu \qquad\qquad (\mu = 1, 2, \cdots, m)$$

respectively. The matrices Δ, Δ' are of course written as

$$\Delta = (\delta_{\mu\nu} - c_{\mu\nu}) , \qquad \Delta' = (\delta_{\nu\mu} - c_{\nu\mu})$$

where $\delta_{\mu\nu} = 0$ for $\mu \neq \nu$, and $\delta_{\mu\nu} = 1$ for $\mu = \nu$. For the case when det Δ = det $\Delta' = 0$, the following facts are known: The associated systems of linear homogeneous equations

(28.8'') $$\rho_\mu - \sum_{\nu=1}^{m} c_{\mu\nu}\rho_\nu = 0 \qquad (\mu = 1, 2, \cdots, m)$$

and

(28.9'') $$\rho'_\mu - \sum_{\nu=1}^{m} c_{\nu\mu}\rho'_\nu = 0 \qquad (\mu = 1, 2, \cdots, m)$$

admit a number r $(r \geqq 1)$ of linearly independent solutions

$$\rho(1) = (\rho_{11}, \ \rho_{12}, \ \cdots, \ \rho_{1m}), \ \cdots,$$
$$\rho(r) = (\rho_{r1}, \ \rho_{r2}, \ \cdots, \ \rho_{rm})$$

and

$$\rho'(1) = (\rho'_{11}, \ \rho'_{12}, \ \cdots, \ \rho'_{1m}), \ \cdots,$$
$$\rho'(r) = (\rho'_{r1}, \ \rho'_{r2}, \ \cdots, \ \rho'_{rm})$$

respectively. The inhomogeneous system (28.8') admits a solution for given $f_1, f_2, \cdots, f_m$ if and only if

(28.12) $$\sum_{\mu=1}^{m} f_\mu \rho'_{j\mu} = 0 \qquad (j = 1, 2, \cdots, m)$$

in other words, for the general solution of (28.9''),

$$\sum_{j=1}^{r} c_j \rho'(j) = \left(\sum_{j=1}^{r} c_j \rho'_{j1}, \ \sum_{j=1}^{r} c_j \rho'_{j2}, \ \cdots, \ \sum_{j=1}^{r} c_j \rho'_{jm} \right)$$

which contains a number r of arbitrary constants $c_1, c_2, \cdots, c_r$, there hold the following relations

(28.12') $$\sum_{\mu=1}^{m} f_\mu \left(\sum_{j=1}^{r} c_j \rho'_{j\mu} \right) = 0$$

If the condition (28.12') is satisfied, then the general solution of (28.8') is given by the sum of a particular solution $\bar{\rho} = (\bar{\rho}_1, \bar{\rho}_2, \cdots, \bar{\rho}_m)$ of (28.8') and the general solution $\sum_{j=1}^{r} c_j\rho(j)$ of (28.8''), that is, by the following expression containing r arbitrary constants $c_1, c_2, \cdots, c_r$,

(28.13) $$\rho = \bar{\rho} + \sum_{j=1}^{r} c_j \rho(j)$$

$$= \left(\overline{\rho}_1 + \sum_{j=1}^{r} c_j \rho_{j1}, \ \overline{\rho}_2 + \sum_{j=1}^{r} c_j \rho_{j2}, \ \cdots, \ \overline{\rho}_m + \sum_{j=1}^{r} c_j \rho_{jm} \right)$$

Similarly, the equations (28.9′) admit a solution for given $g_1, g_2, \cdots, g_m$ if and only if the following relations

$$(28.14) \qquad \sum_{\mu=1}^{m} g_\mu \left(\sum_{j=1}^{r} c_j \rho_{j\mu} \right) = 0$$

hold; and, under the condition (28.14), the general solution of (28.9′) is given by the sum of a particular solution $\overline{\rho}' = (\overline{\rho}'_1, \ \overline{\rho}'_2, \ \cdots, \ \overline{\rho}'_m)$ of (28.9′) and the general solution $\sum_{j=1}^{r} c_j \rho'(j)$ of (28.9″), that is, by the following expression containing r arbitrary constants $c_1, c_2, \cdots, c_r$,

$$(28.15) \qquad \rho' = \overline{\rho}' + \sum_{j=1}^{r} c_j \rho'(j)$$

$$= \left(\overline{\rho}'_1 + \sum_{j=1}^{r} c_j \rho'_{j1}, \ \overline{\rho}'_2 + \sum_{j=1}^{r} c_j \rho'_{j2}, \ \cdots, \ \overline{\rho}'_m + \sum_{j-1}^{r} c_j \rho'_{jm} \right)$$

Accordingly, substituting the solution ρ given by (28.13), if any, in (28.6), we obtain the general solution $\varphi(s)$ of the equation (27.1). The function $\varphi(s)$ contains r arbitrary constants. In fact, if

$$0 = \sum_{\nu=1}^{m} \left(\alpha_\nu(s) + \int_a^b \Gamma_1(s, \ r) \alpha_\nu(r) dr \right) \left(\sum_{j=1}^{r} c_j \rho_{j\nu} \right)$$

then, by the linear independence of (28.5),

$$0 = \sum_{j=1}^{r} c_j \rho_{j\nu} \qquad\qquad (\nu = 1, 2, \cdots, m)$$

This contradicts the fact that $\rho(1), \ \rho(2), \cdots, \ \rho(r)$ are linearly independent solutions of (28.8″). We can also obtain, substituting (28.15) in (28.10), the general solution $\psi(t)$ of (27.2) which contains a number r of arbitrary constants.

Finally, we shall reduce the solvability condition (28.12) to a more readable and usual form as follows.

The term on the left side of (28.12) is, by (28.8) and (28.8′),

$$\sum_{\mu=1}^{m} f_\mu \rho'_{j\mu} = \int_a^b \left[\sum_{\mu=1}^{m} \rho'_{j\mu} \left(\beta_\mu(t) + \int_a^b \Gamma_1(r, \ t) \beta_\mu(r) \, dr \right) \right] f(t) \, dt$$

From (28.10), it is easily seen that the function in brackets on the right side is a solution of (27.2) with $g(t) \equiv 0$, that is, of

$$(28.16) \qquad \psi(t) - \int_a^b K(s, \ t) \psi(s) \, ds = 0$$

On the other hand, the general solution of (28.16) is given by linear combinations of the functions in brackets. Therefore, (28.12) is equivalent to the following: for every solution $\psi(t)$ of (28.16),

$$(28.17) \qquad \int_a^b f(t)\psi(t)\,dt = 0$$

Similarly, we see that the condition (28.14) is equivalent to the following: for every solution $\varphi(s)$ of the equation

$$(28.18) \qquad \varphi(s) - \int_a^b K(s, t)\varphi(t)\,dt = 0\ ,$$

$$\int_a^b g(s)\varphi(s)\,ds = 0$$

29. Fredholm's alternative theorem

The results obtained in Parts 27 and 28 are known as *Fredholm's alternative theorem* concerning a continuous kernel $K(s, t)$ on the domain $a \leq s \leq b,\ a \leq t \leq b$. The theorem reads as follows.

THEOREM 29.1. Either the integral equation of the second kind,

$$(29.1) \qquad f(s) = \varphi(s) - \lambda \int_a^b K(s, t)\varphi(t)\,dt$$

with fixed λ, admits a unique continuous solution $\varphi(s)$ for any continuous function $f(s)$, in particular $\varphi(s) \equiv 0$ for $f(s) \equiv 0$; or the associated homogeneous equation

$$(29.2) \qquad \bar{\varphi}(s) = \lambda \int_a^b K(s, t)\bar{\varphi}(t)\,dt$$

admits a number r $(r \geq 1)$ of linearly independent continuous solutions $\bar{\varphi}_1(s),\ \bar{\varphi}_2(s),\ \cdots,\ \bar{\varphi}_r(s)$. In the first case, the conjugate equation

$$(29.3) \qquad g(s) = \psi(s) - \lambda \int_a^b K(t, s)\psi(t)\,dt$$

also admits a unique continuous solution $\psi(t)$ for any continuous function $g(s)$. In the second case, the associated homogeneous equation

$$(29.4) \qquad \bar{\psi}(s) = \lambda \int_a^b K(t, s)\,\bar{\psi}(t)\,dt$$

admits a number r of linearly independent continuous solutions $\bar{\psi}_1(s),\ \bar{\psi}_2(s),\ \cdots,\ \bar{\psi}_r(s)$. In the second case, the equation (29.1) admits a solution if and only if

$$(29.5) \qquad \int_a^b f(s)\bar{\psi}_j(s)\, ds = 0 \qquad\qquad (j = 1, 2, \cdots, r)$$

If condition (29.5) is satisfied, the general solution of (29.1) is written as

$$(29.6) \qquad \varphi(s) = \varphi^{(1)}(s) + \sum_{j=1}^r c_j \bar{\varphi}_j(s)$$

by means of a particular solution $\varphi^{(1)}(s)$ of (29.1) and r arbitrary constants $c_1, c_2, \cdots, c_r$. Similarly, the conjugate equation (29.2) admits a solution if and only if

$$(29.7) \qquad \int_a^b g(s)\bar{\varphi}_j(s)\, ds = 0 \qquad\qquad (j = 1, 2, \cdots, r)$$

If condition (29.7) is satisfied, the general solution of (29.2) is written as

$$(29.8) \qquad \psi(s) = \psi^{(1)}(s) + \sum_{j=1}^r c_j \bar{\psi}_j(s)$$

by means of a particular solution $\psi^{(1)}(s)$ of (29.2) and r arbitrary constants $c_1, c_2, \cdots, c_r$.

REMARK. This theorem is a generalization of Theorems 23.1 and 23.2 to the case of arbitrary unsymmetric kernels. The theorem also shows that the unique solution of (29.1) exists for any continuous function $f(s)$ if and only if λ is not an eigenvalue.

The kernel without eigenvalues. The unsymmetric kernel

$$(29.9) \qquad K(s, t) = \sum_{\nu=1}^\infty (\nu)^{-2} \sin \nu s \cdot \sin (\nu + 1)t$$

defined on the domain $0 \leq s \leq 2\pi$, $0 \leq t \leq 2\pi$, has no eigenvalue. In fact, since the series (29.9) converges uniformly, we obtain by term-by-term integration that the iterated kernel $K^{(n)}(s, t)$ is given by

$$K^{(n)}(s, t) = \sum_{\nu=1}^\infty \pi^{n-1}[\nu^2(\nu + 1)^2 \cdots (\nu + n - 1)^2]^{-1} \sin \nu s \cdot \sin (\nu + n)t$$

Hence, the Neumann series

$$\sum_{r=1}^\infty \lambda^n K^{(n)}(s, t) = \Gamma(s, t; \lambda)$$

converges uniformly on the domain $0 \leq s \leq 2\pi$, $0 \leq t \leq 2\pi$, for any λ. Therefore, for any λ, the equation

$$f(s) = \varphi(s) - \lambda \int_a^b K(s, t)\varphi(t)\, dt$$

always possesses a unique solution

$$\varphi(s) = f(s) + \int_a^b \Gamma(s, t; \lambda)f(t)\, dt$$

This means that the kernel $K(s, t)$ has no eigenvalues.

REMARK. The proof of Theorem 20.2, for a symmetric kernel, is essentially based upon the relation (20.11) which is derived from the symmetry property of the kernel.

§ 2. The Schmidt expansion theorem and the Mercer expansion theorem

In this section we shall be chiefly concerned with the expansion theorem for arbitrary kernels. As was shown in Part 29, unsymmetric kernels, contrary to symmetric kernels, have, in general, no eigenvalues. The formulation of the expansion theorem in Part 21 is no longer adequate. We shall first show Schmidt's method concerning the expansion theorem for an arbitrary kernel. For the sake of convenience we prepare in Part 30 certain operator-theoretical notations.

30. *Operator-theoretical notations*

Let $K(s, t)$ be a complex-valued continuous function on the domain $a \leqq s \leqq b$, $a \leqq t \leqq b$. We denote by K, as in Part 19, an operator

$$(30.1) \qquad (Kf)(s) = \int_a^b K(s, t)f(t)\, dt$$

which transforms every continuous function $f(s)$ on the interval $a \leqq s \leqq b$ into a continuous function $(Kf)(s)$. Let $L(s, t)$ be a complex-valued continuous function. Then the kernel $M(s, t)$ defined by

$$(30.2) \qquad M(s, t) = \int_a^b K(s, r)L(r, t)\, dr$$

is said to be the *composition of the kernels* $K(s, t)$ and $L(s, t)$. Clearly, the operator M defined by $M(s, t)$ satisfies the following relation

$$(30.3) \qquad Mf = K(Lf), \quad \text{i.e., } M = KL$$

As in Part 20, we define the inner product of a pair of continuous functions $f(s)$, $g(s)$ and the norm of $f(s)$ by

$$(30.4) \qquad (f,\, g) = \int_a^b f(s)\overline{g(s)}\, ds\,, \qquad ||f|| = (f,f)^{\frac{1}{2}}$$

respectively. Further, we define the *transposed conjugate kernel* $K^*(s, t)$ of $K(s, t)$ by

$$(30.5) \qquad\qquad K^*(s,\, t) \equiv \overline{K(t,\, s)}$$

Then, since

$$(Kf,\, g) = \int_a^b \left[\int_a^b K(s,\, t)\, f(t)\, dt \right] \overline{g(s)}\, ds$$
$$= \int_a^b f(t) \left[\overline{\int_a^b K^*(t,\, s)\, g(s)\, ds} \right] dt$$

we have for any pair of continuous functions $f(s)$, $g(s)$, the following relation

$$(30.6) \qquad\qquad (Kf,\, g) = (f,\, K^*g)$$

In view of (30.6) the Hermitian symmetry property

$$(30.7) \qquad\qquad K(s,\, t) \equiv K^*(s,\, t)$$

is equivalent to

$$(30.8) \qquad\qquad (Kf,\, g) = (f,\, Kg)$$

By making use of (30.6), we have further

$$(KLf,\, g) = (K(Lf),\, g) = (Lf,\, K^*g) = (f,\, L^*(K^*g))$$
$$= (f,\, L^*K^*g)$$

Hence we obtain the relation

$$(30.9) \qquad\qquad (KL)^* = L^*K^*$$

which plays an important role in the sequel. We note here that the operators

$$(30.10) \qquad\qquad KK^* \quad \text{and} \quad K^*K$$

are both Hermitian symmetric, for we see easily that

$$(30.11) \qquad\qquad K^{**} = (K^*)^* = K$$

31. *The Schmidt expansion theorem*

We start with the following theorem.

THEOREM 31.1. Let $K(x, y) \not\equiv 0$, then the Hermitian symmetric operators

$$(31.1) \qquad\qquad A = KK^* \quad \text{and} \quad B = K^*K$$

have the same eigenvalues with the same multiplicity. Further, their eigenvalues are all positive.

Proof. We shall prove first that every eigenvalue of A (B) is positive. Since

$$(KK^*)(s,\ s) = \int_a^b K(s, r)\overline{K(s, r)}\, dr = \int_a^b |\, K(s, r)\,|^2\, dr$$

we see that $(KK^*)(s, t) \equiv A(s, t) \not\equiv 0$, and similarly $B(s, t) \not\equiv 0$. Both A and B are Hermitian symmetric, hence, according to Theorem 20.2, both operators have eigenvalues. Let λ be an eigenvalue of A, that is,

$$\lambda A\varphi = \varphi\, , \qquad \varphi(s) \not\equiv 0$$

Then we have

(31.2)
$$0 < (\varphi, \varphi) = (\lambda A\varphi, \varphi) = (\lambda KK^*\varphi, \varphi)$$
$$= (\lambda K^*\varphi, K^*\varphi) = \lambda(K^*\varphi, K^*\varphi)$$

while

$$(K^*\varphi, K^*\varphi) \geqq 0$$

This implies that all the eigenvalues of A are positive. The proof for B is carried out in the same way.

Accordingly, as in Part 21, we write the eigenvalues of A and the corresponding orthonormal system of eigenfunctions as follows:

(31.3)
$$\lambda_1^2 \leqq \lambda_2^2 \leqq \cdots \leqq \lambda_j^2 \leqq \cdots \qquad (\lambda_j > 0)$$
$$\varphi_1, \varphi_2, \cdots, \varphi_j, \cdots \qquad (\lambda_j^2 A\varphi_j = \varphi_j)$$

We now set

(31.4)
$$\lambda_j K^*\varphi_j = \psi_j$$

Then, we have

$$\lambda_j^2 B\psi_j = \lambda_j^2 K^*K\psi_j = \lambda_j^2 K^*K(\lambda_j K^*\varphi_j)$$
$$= \lambda_j K^*\lambda_j^2 KK^*\varphi_j = \lambda_j K^*\lambda_j^2 A\varphi_j = \lambda_j K^*\varphi_j = \psi_j$$

and hence

(31.5)
$$\lambda_j^2 B\psi_j = \psi_j$$

Thus λ_j^2 is also an eigenvalue of B, and ψ_j the corresponding eigenfunction. Further, by virtue of the relation $(\varphi_i, \varphi_j) = \delta_{ij}$ and (30.6), we have

$$(\psi_i, \ \psi_j) = (\lambda_i K^* \varphi_i, \ \lambda_j K^* \varphi_j) = (\lambda_i \lambda_j K K^* \varphi_i, \ \varphi_j)$$

$$= \frac{\lambda_j}{\lambda_i} \lambda_i^2 (A\varphi_i, \ \varphi_j) = \frac{\lambda_j}{\lambda_i}(\varphi_i, \ \varphi_j) = \frac{\lambda_j}{\lambda_i} \delta_{ij}$$

hence we see that the eigenfunctions $\{\psi_j\}$ of B satisfy the orthonormality relations. Consequently, it only remains to prove that $\{\psi_j\}$ exhausts all eigenfunctions of B. To prove this, it is sufficient to show that

(31.6)   if $\lambda^2 B\psi = \psi$, $\psi(s) \not\equiv 0$, then

$$\lambda K\psi = \varphi \text{ satisfies } \lambda^2 A\varphi = \varphi .$$

In fact, if (31.6) is true, then φ is written as a linear combination of finite number of φ_j with $\lambda_j^2 = \lambda^2$. On the other hand,

$$\psi = \lambda^2 B\psi = \lambda K^* \lambda K\psi = \lambda K^* \varphi$$

Hence ψ is written as a linear combination of finite number of $\psi_j = \lambda_j K^* \varphi_j$. Accordingly, the eigenvalues of B and the corresponding orthonormal system of eigenfunctions are given by

(31.7) $\lambda_1^2 \leqq \lambda_2^2 \leqq \cdots \leqq \lambda_j^2 \leqq \cdots$

$$\psi_1 = \lambda_1 K^* \varphi_1, \ \psi_2 = \lambda_2 K^* \varphi_2, \cdots, \ \psi_j = \lambda_j K^* \varphi_j, \cdots$$

We now prove (31.6) by means of the following calculation

$$\lambda^2 A\varphi = \lambda^2 K K^* \varphi = \lambda^2 K K^* (\lambda K\psi) = \lambda K \lambda^2 K^* K\psi$$

$$= \lambda K \lambda^2 B\psi = \lambda K\psi = \varphi$$

Thus the proof is completed, q.e.d.

The Schmidt expansion theorem reads as follows.

THEOREM 31.2. Every continouus function of the form $(Kf)(s)$, where $f(s)$ is a continuous function, can be expanded in Fourier series with respect to $\{\varphi_j\}$, which converges absolutely and uniformly on the interval $a \leqq s \leqq b$. Every continuous function of the form $(K^*f)(s)$ also can be expanded in Fourier series with respect to $\{\psi_j\}$, which converges absolutely and uniformly on the interval $a \leqq s \leqq b$.

Proof. Denoting by (Kf, φ_j) the expansion coefficient of $(Kf)(s)$ with respect to $\{\varphi_j\}$, we have, by (31.4),

$$(Kf, \ \varphi_j) = (Kf, \ \lambda_j^2 A\varphi_j) = (Kf, \ \lambda_j^2 K K^* \varphi_j)$$

$$= (K^* Kf, \ \lambda_j^2 K^* \varphi_j) = \lambda_j(Bf, \ \psi_j) = \lambda_j(f, \ B\psi_j)$$

$$= \lambda_j^{-1}(f, \ \psi_j)$$

By the Schwarz inequality,

$$\left(\sum_{j=n}^{m}\left|(f,\ \psi_j)\frac{\varphi_j(s)}{\lambda_j}\right|\right)^2 \leqq \sum_{j=n}^{m}|(f,\ \psi_j)|^2 \sum_{j=n}^{m}\left|\frac{\varphi_j(s)}{\lambda_j}\right|^2$$

By the Bessel inequality and (31.6), we have further

$$\int_a^b |K(s,\ t)|^2\, dt \geqq \sum_{j=1}^{\infty}\left|\frac{\varphi_j(s)}{\lambda_j}\right|^2,\quad \|f\|^2 \geqq \sum_{j=1}^{\infty}|(f,\ \psi_j)|^2$$

Therefore the series $\sum_{j=1}^{\infty}\lambda_j^{-1}(f,\ \psi_j)\varphi_j(s)$, and hence the Fouries expansion of $(Kf)(s)$, converges absolutely and uniformly on the interval $a \leqq s \leqq b$.

We now set

(31.8) $$\Theta(s) = (Kf)(s) - \sum_{j=1}^{\infty}(f,\ \psi_j)\frac{\varphi_j(s)}{\lambda_j}$$

Then it remains to prove that $\Theta(s) \equiv 0$. Since

(31.9) $$(\Theta,\ \varphi_j) = (Kf,\ \varphi_j) - (f,\ \psi_j)\frac{1}{\lambda_j} = 0\ ,$$

Θ is orthogonal to every φ_j. Further, we have

$$(K^*\Theta,\ K^*\Theta) = (KK^*\Theta,\ \Theta) = (A\Theta,\ \Theta)$$

On the other hand, according to Theorem 21.4, $(A\Theta)(s)$ can be expanded in Fourier series with respect to $\{\varphi_j\}$, which converges uniformly; hence, by term-by-term integration and (31.9), we see that $(A\Theta,\ \Theta) = 0$. Therefore the continuous function $(K^*\Theta)(s)$ must be identically zero. Hence, by virtue of (31.8) and (31.9), we obtain, by term-by-term integration, that

$$0 = (K^*\Theta,\ f) = (\Theta,\ Kf) = \left(\Theta,\ \left(\Theta + \sum_{j=1}^{\infty}(f,\ \psi_j)\frac{\varphi_j}{\lambda_j}\right)\right)$$
$$= (\Theta,\ \Theta)$$

Since $\Theta(s)$ is continuous, $\Theta(s)$ must be identically zero, that is $\Theta(s) \equiv 0$. The proof for $(K^*f)(s)$ is carried out in an analogous manner.

32. Application to Fredholm integral equation of the first kind

An equation, in the unknown $\varphi(s)$,

(32.1) $$f(s) = \int_a^b K(s,\ t)\varphi(t)\, dt$$

with a continuous kernel $K(s,\ t)$, $f(s)$ being given and continuous, is known as *Fredholm integral equation of the first kind*.

THEOREM 32.1. If the equation (32.1) admits a continuous solution $\varphi(s)$ for given $f(s)$, then $f(s)$ can be expanded in a Fourier series

$$(32.2) \qquad f(s) = \sum_{j=1}^{\infty} \beta_j \varphi_j(s)$$

with respect to the orthonormal system of the eigenfunctions $\{\varphi_j\}$ of $A = KK^*$, which converges absolutely and uniformly on the interval $a \leq s \leq b$. Conversely, if $f(s)$ is of the form (32.2), and if the series

$$(32.3) \qquad \sum_{j=1}^{\infty} \beta_j \lambda_j \psi_j(s)$$

where ψ_j are the eigenfunctions of $B = K^*K$ so that $B\psi_j = \lambda_j^{-2}\psi_j$, converges absolutely and uniformly on the interval $a \leq s \leq b$, then the equation (32.1) admits a solution $\varphi(s)$ which is given by (32.3).

Proof. Proceeds similarly as in Part 31.

REMARK. According to the theorem, we see that, in general, the equation (32.1) can not be solved. Furthermore, the uniqueness of the solution $\varphi(s)$, if any, is equivalent to the following property of the kernel: if a continuous function $g(s)$ satisfies

$$(32.4) \qquad \int_a^b K(s,\ t)g(t)\,dt \equiv 0$$

then $g(s) \equiv 0$. A kernel $K(s, t)$ which possesses such a property is called a *closed kernel*. A continuous closed kernel is $\equiv 0$.

We can derive a necessary and sufficient condition for the existence of the solution of (32.1)—Picard's Theorem—by making use of the concept of Lebesgue integral. The reader is referred, for example, to T. Lalésco, *Introduction à la théorie des équations intégrales*, Paris, 1912, p. 99.

33. *Positive definite kernels. Mercer's expansion theorem*

A complex-valued continuous kernel $K(s, t)$ defined on a domain $a \leq s \leq b$, $a \leq t \leq b$ is called a *positive definite kernel*, when $K(s, t)$ satisfies

$$(33.1) \qquad \int_a^b \int_a^b K(s,\ t)f(s)\overline{f(t)}\,ds\,dt = (Kf,\ f) \geq 0$$

for any continuous function $f(s)$. For example, $A(s, t)$ and $B(s, t)$ defined in Part 31 are both positive definite.

THEOREM 33.1. In order that $K(s, t)$ is positive definite, it is necessary and sufficient that, for any finite number of arbitrary points $\{x_j\}$, $a < x_1 < \cdots < x_n < b$, and arbitrary complex numbers $\xi_1, \xi_2, \cdots, \xi_n$, the following relation holds

$$(33.2) \qquad \sum_{i,j=1}^{n} K(x_i, \ x_j) \xi_i \bar{\xi}_j \geqq 0$$

Proof.[1] To prove sufficiency, we set

$$x_i = a + \frac{i}{n+1}(b - a)$$

and we see that the left side of (33.1) equals

$$\lim_{n \to \infty} \frac{(b-a)^2}{(n+1)^2} \sum_{i,j=1}^{n} K(x_i, \ x_j) f(x_i) \overline{f(x_j)}$$

To prove the necessity of the conditions, we proceed as follows: Suppose that there exist $\{x_j\}$ and $\{\xi_j\}$ for which the left side of (33.2) is equal to $-\alpha < 0$. Then we shall prove that there exists a continuous function $f(s)$ for which (33.1) does not hold. Let $a < c < b$. Then, for given c, we can choose sufficiently small positive numbers ε, η such that $a < c - \varepsilon - \eta < c + \varepsilon + \eta < b$. For such c, ε and η, we define $\theta_{\varepsilon,\eta}(s, c)$ as follows:

$$\theta_{\varepsilon,\eta}(s, \ c) = \begin{cases} 0, \ a \leqq s \leqq c - \varepsilon - \eta, \ c + \varepsilon + \eta \leqq s \leqq b \\ 1, \ c - \eta \leqq s \leqq c + \eta \end{cases}$$

and in the rest of the interval $[a, b]$, $\theta_{\varepsilon,\eta}(s, c)$ is linear in s, so that its graph is as in Fig. 1. Clearly, $\theta_{\varepsilon,\eta}(s, c)$ is a continuous function

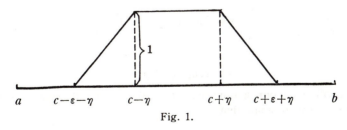

Fig. 1.

of s on the interval $[a, b]$. Now we choose $\varepsilon > 0$, and $\eta > 0$ so small that no pair of the intervals

$$(-\varepsilon - \eta + x_i, \ x_i + \varepsilon + \eta)$$

[1] The following proof is adapted from that given in Widder, D. V., *Laplace Transform*, Princeton, 1946, p. 271.

have a common point and $a < x_1 - \varepsilon - \eta$, $x_n + \varepsilon + \eta < b$. For such ε and η, we define $\theta_{\varepsilon,\eta}(s, x_i)$ as in above, and we set

$$\theta(s) = \sum_{i=1}^{n} \xi_i \theta_{\varepsilon,\eta}(s, x_i)$$

Set further

$$L_n(x, y) = \sum_{i,j=1}^{n} K(x_i + x, y_j + y)\xi_i\bar{\xi}_j$$

Then, according to our construction, we see easily that

$$(33.3) \qquad (K\theta, \theta) = \int_{-\eta}^{\eta}\int_{-\eta}^{\eta} L_n(x, y) \, dx \, dy$$
$$+ \sum_{i,j=1}^{n} \iint_{q_{ij}} K(x, y)\theta(x)\theta(y) \, dx \, dy$$

where q_{ij} is the region between the square

$$x_i - \varepsilon - \eta \leq x \leq x_i + \varepsilon + \eta, \ x_j - \varepsilon - \eta \leq y \leq x_j + \varepsilon + \eta$$

and the square

$$x_i - \eta \leq x \leq x_i + \eta, \ x_j - \eta \leq y \leq x_j + \eta$$

Furthermore, for any point (x, y) in the region q_{ij},

$$| \theta(x)\overline{\theta(y)} | \leq | \xi_i\bar{\xi}_j |$$

hence, the absolute value of the second term on the right side of (33.3) is not greater than

$$M4\varepsilon(2\eta + \varepsilon)\left(\sum_{i=1}^{n} | \xi_i |\right)^2$$

where

$$M = \sup_{(x,y)} | K(x, y) |$$

According to our assumption, $L_n(0, 0) = -\alpha < 0$; hence we can choose $\eta > 0$ so small that

$$L_n(x, y) < -\frac{1}{2}\alpha$$

whenever $| x | < \eta$, $| y | < \eta$.

Then, for such an η, we have

$$\int_{-\eta}^{\eta} \int_{-\eta}^{\eta} L_n(x, y)\, dx\, dy < -2\alpha\eta^2$$

and hence we obtain

$$(K\theta, \theta) < -2\alpha\eta^2 + M4\varepsilon(2\eta + \varepsilon)\left(\sum_{i=1}^{n} |\xi_i|\right)^2$$

Since the right side of the above inequality tends to $-2\alpha\eta^2$ as $\varepsilon \to 0$, we can choose η and ε sufficiently small so that, by setting $\theta(s) = f(s)$, $(Kf, f) < 0$ holds. This is a contradiction, q.e.d.

THEOREM 33.2. If a kernel $K(s, t)$ is positive definite, then $K(s, t)$ is Hermitian symmetric and $K(s, s) \geqq 0$ for any s in $[a, b]$.

Proof. By Theorem 33.1, we have, for any point x_0 in (a, b),

$$K(x_0, x_0)\xi_0\bar{\xi}_0 \geqq 0$$

hence

$$K(x_0, x_0) \geqq 0$$

Therefore, because of the continuity of $K(s, t)$, we obtain that

(33.4) $$K(s, s) \geqq 0$$

for any point s in $[a, b]$.

By Theorem 33.1, for any pair of points x_0, x_1 in (a, b), we have

$$K(x_0, x_0)\xi_0\bar{\xi}_0 + K(x_1, x_1)\xi_1\bar{\xi}_1 + K(x_0, x_1)\xi_0\bar{\xi}_1$$
$$+ K(x_1, x_0)\xi_1\bar{\xi}_0 \geqq 0$$

Hence, by (33.4), we have

$$K(x_0, x_1)\xi_0\bar{\xi}_1 + K(x_1, x_0)\xi_1\bar{\xi}_0 = \text{real number}$$

Then, setting $\xi_0 = \xi_1 = 1$ and $\xi_0 = 1$, $\xi_1 = \sqrt{-1}$, we obtain

$$K(x_0, x_1) + K(x_1, x_0) = \text{real number}$$

and

$$-K(x_0, x_1)\sqrt{-1} + K(x_1, x_0)\sqrt{-1} = \text{real number}$$

respectively. Hence we see that $K(x_0, x_1) = \overline{K(x, x_0)}$ for any x_0, x_1 in (a, b). Thus, because of the continuity of $K(s, t)$, we obtain that

(33.5) $$K(s, t) \equiv \overline{K(t, s)}$$

for any s, t in $[a, b]$, q.e.d.

Since it is Hermitian symmetric, a positive definite kernel $K(s, t) \not\equiv 0$

has an eigenvalue. Furthermore its eigenvalues are all positive. In fact, if

$$\lambda K\varphi = \varphi , \quad \varphi(s) \not\equiv 0$$

then

$$0 < (\varphi, \varphi) = (\lambda K\varphi, \varphi)$$

while $(K\varphi, \varphi) \geq 0$. Accordingly, all the eigenvalues of $K(s, t)$ and the corresponding orthonormal system of the eigenfunctions may be written as follows:

$$0 < \lambda_1 \leq \lambda_2 \leq \cdots \leq \lambda_j \leq \cdots ;$$

$$\varphi_1, \varphi_2, \cdots, \varphi_j, \cdots; \quad \lambda_j K\varphi_j = \varphi_j$$

We may now formulate the *Mercer expansion theorem*.

THEOREM 33.3. A positive definite kernel $K(s, t)$ can be expanded in a series

$$(33.6) \qquad K(s, t) \equiv \sum_{j=1}^{\infty} \lambda_j^{-1} \varphi_j(s) \overline{\varphi_j(t)}$$

which converges absolutely and uniformly on the domain $a \leq s \leq b$, $a \leq t \leq b$.

Proof. The Fourier expansion of $(Kf)(s)$ is given by

$$\sum_{j=1}^{\infty} (Kf, \varphi_j)\varphi_j(s) = \sum_{j=1}^{\infty} (f, K\varphi_j)\varphi_j(s)$$

$$= \sum_{j=1}^{\infty} \lambda_j^{-1}(f, \varphi_j)\varphi_j(s)$$

Hence, setting

$$K_n(s, t) = K(s, t) - \sum_{j=1}^{n} \lambda_j^{-1}\varphi_j(s)\overline{\varphi_j(t)}$$

for each n, we have

$$(K_n f)(s) = \sum_{j=n+1}^{\infty} \lambda_j^{-1}\varphi_j(s)(f, \varphi_j)$$

Since λ_j is positive, we have

$$(K_n f, f) = \sum_{j=n+1}^{\infty} \lambda_j^{-1}\overline{(f, \varphi_j)}(f, \varphi_j) \geq 0$$

which shows that $K_n(s, t)$ is positive definite. Accordingly, Theorem 33.2 implies that

$$K_n(s, \ s) = K(s, \ s) - \sum_{j=1}^{n} \lambda_j^{-1}\varphi_j(s)\overline{\varphi_j(s)} \geqq 0$$

Hence the series

$$\sum_{j=1}^{\infty} \lambda_j^{-1} \mid \varphi_j(s) \mid^2$$

converges and is not greater than $K(s, s)$.

Now we set

(33.7) $$S(s, \ t) = \sum_{j=1}^{\infty} \lambda_j^{-1}\varphi_j(s)\overline{\varphi_j(t)}$$

Then, we obtain, by the Schwarz and the Bessel inequalities, that

$$\left(\sum_{j=n}^{m} \mid \lambda_j^{-1}\varphi_j(s)\overline{\varphi_j(t)} \mid \right)^2 \leqq \sum_{j=n}^{m} \lambda_j^{-1} \mid \varphi_j(s) \mid^2 \sum_{j=n}^{m} \lambda_j^{-1} \mid \varphi_j(t) \mid^2$$

$$\leqq \sum_{j=n}^{m} \lambda_j^{-1} \mid \varphi_j(s) \mid^2 K(t, \ t)$$

Hence, the series $S(s, t)$ converges absolutely and uniformly with respect to t for s fixed and also with respect to s for t fixed.

We now set

$$R(s, \ t) = K(s, \ t) - S(s, \ t)$$

Then, for any continuous function $f(s)$, we have

$(*)$ $$\int_a^b K(s, \ t)f(t) \, dt = \int_a^b S(s, \ t)f(t) \, dt + \int_a^b R(s, \ t)f(t) \, dt$$

According to the Hilbert-Schmidt expansion theorem, the term on the left side can be expanded in a series

$$\sum_{j=1}^{\infty} (Kf, \ \varphi_j)\varphi_j(s) = \sum_{j=1}^{\infty} (f, \ K\varphi_j)\varphi_j(s)$$

$$= \sum_{j=1}^{\infty} \lambda_j^{-1}(f, \ \varphi_j)\varphi_j(s)$$

On the other hand, since the series $S(s, t)$ converges uniformly with respect to t, we obtain, by term-by-term integration, that the first term on the right side of $(*)$ is equal to $\sum_{j=1}^{\infty} \lambda_j^{-1}(f, \varphi)\varphi_j(s)$. Hence we obtain

(33.8) $$\int_a^b R(s, \ t)f(t) \, dt \equiv 0$$

Since the series $S(s, t)$ converges uniformly with respect to t for s fixed, $R(s, t)$ is a continuous function of t for any fixed s. Hence (33.8) implies, by setting $f(t) = \overline{R(s, t)}$ in (33.8), that $R(s, t)$ is identically zero as a function of t for any fixed s; hence $R(s, t) \equiv 0$. Thus we obtain

$$K(s, t) \equiv \sum_{j=1}^{\infty} \lambda_j^{-1} \varphi_j(s) \overline{\varphi_j(t)}$$

We shall prove that the series on the right side converges uniformly. We recall first that the series $\sum_{j=1}^{\infty} \lambda_j^{-1} \mid \varphi_j(s) \mid^2$ of positive continuous functions is equal to the continuous function $K(s, s)$. By virtue of the above fact, we can find, for any $\varepsilon > 0$ and any given s_0, $n = n(s_0)$, such that

$$\varepsilon > K(s_0, s_0) - \sum_{j=1}^{n} \lambda_j^{-1} \mid \varphi_j(s_0) \mid^2 \geqq 0$$

Furthermore, since $K(s, s)$ and $\sum_{j=1}^{n} \lambda_j^{-1} \mid \varphi_j(s) \mid^2$ are both continuous, we can find an open set $U(s_0)$ containing s_0 such that

(33.9) $$2\varepsilon \geqq K(s, s) - \sum_{j=1}^{n(s_0)} \lambda_j^{-1} \mid \varphi_j(s) \mid^2 \geqq 0$$

whenever $s \in U(s_0)$.

On the other hand, since it is bounded and closed, the interval $[a, b]$ is covered completely by a finite number of open sets $U(s_1)$, $U(s_2), \cdots, U(s_k)$, in other words, every point s in $[a, b]$ is contained in some $U(s_j)$. Now, we set

$$n_0 = \max_{1 \leqq j \leqq k} n(s_j)$$

Then, the fact that every term of $\sum_{j=1}^{\infty} \lambda_j^{-1} \mid \varphi_j(s) \mid^2$ is positive, together with (33.9), implies that

(33.10)   for every $s \in [a, b]$

$$2\varepsilon \geqq K(s, s) - \sum_{j=1}^{n} \lambda_j^{-1} \mid \varphi_j(s) \mid^2 \geqq 0$$

whenever $n \geqq n_0$.

Hence, we see that the series $\sum_{j=1}^{\infty} \lambda_j^{-1} \mid \varphi_j(s) \mid^2$ converges absolutely to $K(s, s)$ and uniformly in s. By making use of the Schwarz inequality, we have

$$\left(\sum_{j=n}^{m} \lambda_j^{-1} \mid \varphi_j(s) \overline{\varphi_j(t)} \mid \right)^2 \leqq \sum_{j=n}^{m} \lambda_j^{-1} \mid \varphi_j(s) \mid^2 \sum_{j=n}^{m} \lambda_j^{-1} \mid \varphi_j(t) \mid^2$$

Thus we see that the series on the right side of (33.6) converges absolutely and uniformly for s and t, q.e.d.

REMARK. The Hilbert-Schmidt expansion theorem for a positive definite kernel $K(s, t)$ is easily proved by (33.6).

For any continuous kernel $K(s, t)$, the kernels $A(s, t)$ and $B(s, t)$ defined by (Part 31)

$$A(s, t) = \int_a^b K(s, r)K^*(r, t)\, dr$$

$$B(s, t) = \int_a^b K^*(s, r)K(r, t)\, dr$$

are both positive definite. Hence $A(s, t)$ and $B(s, t)$ can be expanded in absolutely and uniformly convergent series

$$A(s, t) \equiv \sum_{j=1}^\infty \lambda_j^{-2}\varphi_j(s)\overline{\varphi_j(t)}$$

$$B(s, t) \equiv \sum_{j=1}^\infty \lambda_j^{-2}\psi_j(s)\overline{\psi_j(t)}$$

with respect to the eigenfunctions $\{\varphi_j\}$ and $\{\psi_j\}$ as in Part 31.

As a consequence we have the following theorem.

Theorem 33.4. The necessary and sufficient condition that A and B have the same eigenvalues, together with the corresponding eigenfunctions, is that $A = B$.

In general, a kernel $K(s, t)$ which satisfies

(33.11) $A = B$, that is, $KK^* = K^*K$

is called a *normal kernel*. Hermitian symmetric kernels and *Hermitian skew-symmetric kernels* which satisfy

(33.12) $K(s, t) \equiv -\overline{K(t, s)}$

are clearly normal.

THEOREM 33.5. A Hermitian skew-symmetric kernel $K(s, t) \not\equiv 0$, which satisfies (33.12), has eigenvalues. Its eigenvalues are all purely imaginary numbers.

Proof. The truth of the theorem is obvious, since $\sqrt{-1}\cdot K(s, t)$ is Hermitian symmetric, q.e.d.

§3. Singular integral equations

In the preceding paragraphs, we were concerned with the case when the interval $[a, b]$ under consideration is finite and closed and the kernel $K(s, t)$ is continuous for

$$a \leqq s \leqq b, \quad a \leqq t \leqq b$$

Our earlier treatment may no longer be valid, if the kernel $K(s, t)$ is not continuous, or, if the interval $[a, b]$ is not finite, that is, $a = -\infty$, $b = +\infty$, or both $a = -\infty$ and $b = +\infty$. These cases are regarded as the *singular cases*, the corresponding integral equations are called *singular integral equations*, and the kernels *singular kernels*. In what follows we shall make some remarks concerning these cases.

34. *Discontinuous kernels*

In the case when the interval $[a, b]$ is finite and closed, Fredholm alternative theorem sometimes holds, even if the kernel is not continuous. If, for a kernel $K(s, t)$, there exist two sets $\{\alpha_\nu(s)\}$, $\{\beta_\nu(t)\}$ of linearly independent continuous functions such that (28.2) holds, and if, for the kernel

$$(34.1) \qquad K(s, t) - \sum_{\nu=1}^{m} \alpha_\nu(s)\beta_\nu(t) = K_1(s, t)$$

the procedure in Part 27 is valid, then the theorem also holds for the kernel $K(s, t)$.

An example of such kernels is

$$(34.2) \qquad K(s, t) = H(s, t) \mid s - t \mid^{-\alpha}, \quad 0 < \alpha < \frac{1}{2}$$

where $H(s, t)$ is a continuous function in both s and t. For such a kernel $K(s, t)$, we can find a continuous *degenerated kernel* $\sum_{\nu=1}^{m} \alpha_\nu(s)\beta_\nu(t)$ such that

$$(34.3) \qquad \begin{aligned} &\int_a^b \int_a^b \left| K(s, t) - \sum_{\nu=1}^{m} \alpha_\nu(s)\beta_\nu(t) \right|^2 ds\, dt < 1 \\ &\sup_{a \leqq t \leqq b} \int_a^b \left| K(s, t) - \sum_{\nu=1}^{m} \alpha_\nu(s)\beta_\nu(t) \right|^2 ds < \infty \\ &\sup_{a \leqq s \leqq b} \int_a^b \left| K(s, t) - \sum_{\nu=1}^{m} \alpha_\nu(s)\beta_\nu(t) \right|^2 dt < \infty \end{aligned}$$

In fact, it is easily seen, from (34.2), that there exists a continuous kernel $K_1(s, t)$ such that

$$\int_a^b \int_a^b \mid K(s, t) - K_1(s, t) \mid^2 ds\, dt < 1$$

$$\sup_{a \leqq t \leqq b} \int_a^b \mid K(s, t) - K_1(s, t) \mid^2 ds < \infty$$

$$\sup_{a \leqq s \leqq b} \int_a^b \mid K(s, t) - K_1(s, t) \mid^2 dt < \infty$$

Hence, the degenerated kernel for $K_1(s, t)$, defined as in Part 28, satisfies (34.3).

If (34.3) is satisfied, then the procedure in Part 27, and hence that in Part 28, is also valid. Thus Fredholm's theorem also holds.

Boundary value problems for elliptic partial differential equations, such as the *Dirichlet problem* or the *Neumann problem*, can be reduced to an integral equation with a kernel like (34.2), t and s denoting higher-dimensional vectors. Fredholm's study was motivated by the Dirichlet problem. Within the scope of this book, we can not touch on the subject of partial differential equations. For the relation between integral equations and partial differential equaions, we refer the reader to the followings: É. Picard, *Leçons sur quelques problèmes aux limites de la théorie des équations différentielles*. Paris, 1930; R. Courant and D. Hilbert, *Methods of Mathematical Physics*, Vol. 1, New York, 1953; D. Hilbert, *Grundzüge einer allgemeinen Theorie der linearen Integralgleichungen*, Leipzig and Berlin, 1912; and especially for the Dirichlet and the Neumann problem, O. D. Kellog, *Foundations of Potential Theory*, Berlin, 1929.

35. Examples. Band spectrum

We shall give some examples of the singular cases. From the formula

$$\int e^{px} \sin qx \, dx = e^{px} \frac{p \sin qx - q \cos qx}{p^2 + q^2}$$

there follows

$$\int_0^\infty e^{-at} \sin st \, dt = \frac{s}{a^2 + s^2} \qquad (a > 0)$$

From the formula

$$\int_0^\infty \frac{\cos px}{1 + x^2} dx = \frac{\pi}{2} e^{-p} \qquad (p > 0)$$

there follows, by differentiation with respect to p,

$$\int_0^\infty \frac{t}{a^2 + t^2} \sin st \, dt = \frac{\pi}{2} e^{-as} \qquad (a, s > 0)$$

Therefore, we have the integral formula

(35.1) $$\int_0^\infty \sqrt{(2/\pi)} \sin st \left(\sqrt{(\pi/2)} \, e^{-at} + \frac{t}{a^2 + t^2} \right) dt$$

$$= \sqrt{(\pi/2)}\, e^{-as} + \frac{s}{a^2 + s^2} \qquad\qquad (a,\ s > 0)$$

Accordingly, on the infinite open interval $(0, \infty)$, the continuous kernel

$$(35.2) \qquad\qquad K(s,\ t) = \sqrt{(2/\pi)} \sin st \qquad\qquad (0 < s,\ t < \infty)$$

has infinitely many eigenfunctions

$$(35.3) \qquad\qquad \sqrt{(\pi/2)}\, e^{-at} + \frac{t}{a^2 + t^2} \qquad\qquad (a > 0)$$

for the eigenvalue 1. In other words, the multiplicity of the eigenvalue 1 has the same power as the *continuum*. If the interval $[a, b]$ considered is finite and closed, the multiplicity of eigenvalues must be at most finite (Theorem 21.3).

The above example is due to Weyl. We now consider another example given by Picard. For real α, we have

$$\int_{-\infty}^{\infty} e^{-|s-t|}\, e^{i\alpha t}\, dt \qquad\qquad (i = \sqrt{-1})$$

$$= \int_{-\infty}^{s} e^{-(s-t)+i\alpha t}\, dt + \int_{s}^{\infty} e^{-(t-s)+i\alpha t}\, dt$$

$$= \left[\frac{e^{-(s-t)+i\alpha t}}{1 + i\alpha}\right]_{-\infty}^{s} + \left[\frac{e^{-(t-s)+i\alpha t}}{-1 + i\alpha}\right]_{s}^{\infty}$$

$$= \frac{e^{i\alpha s}}{1 + i\alpha} - \frac{e^{i\alpha s}}{-1 + i\alpha} = \frac{2e^{i\alpha s}}{1 + \alpha^2}$$

Accordingly, we obtain

$$(35.4) \qquad\qquad \int_{-\infty}^{\infty} e^{-|s-t|} e^{i\alpha t}\, dt = \left(\frac{2}{1 + \alpha^2}\right) e^{i\alpha s} \qquad\qquad (\alpha = \text{real})$$

Therefore, on the infinite open interval $(-\infty, \infty)$, the kernel

$$(35.5) \qquad\qquad K(s,\ t) = e^{-|s-t|} \qquad\qquad (-\infty < s,\ t < \infty)$$

has eigenfunctions for any

$$\lambda = \frac{1 + \alpha^2}{2} > \frac{1}{2}$$

In other words, every $\lambda > \frac{1}{2}$ is an eigenvalue of the kernel. If the interval $[a, b]$ considered is finite and closed, the continuous kernel has at most denumerably many eigenvalues, and there is no limiting point of eigenvalues except for $\pm\infty$ (Theorem 21.3).

Spectrum. The set of all the eigenvalues is called the *spectrum.* The set of all the isolated points in the spectrum is called the *line spectrum.* An eigenvalue which is not contained in the line spectrum is said to belong to the *band spectrum.*

It is noteworthy that, contrary to what is true for the continuous kernel on a finite closed interval, the singular kernel may admit a band spectrum, as is shown in the example mentioned above.

CHAPTER 4

VOLTERRA INTEGRAL EQUATIONS

§1. Volterra integral equations of the second kind

36. *Existence and uniqueness of solutions*

Let $K(s, t)$ be a complex-valued continuous function defined on a domain $a \leq s \leq b, a \leq t \leq b$. Let $f(s)$ be a complex-valued continuous function defined on the interval $a \leq s \leq b$. We shall be concerned with the problem of finding a continuous function $\varphi(s)$ satisfying the following equation

$$(36.1) \qquad f(s) = \varphi(s) - \lambda \int_a^s K(s, t)\, \varphi(t)\, dt$$

Such an equation is called a *Volterra integral equation of the second kind*.

Suppose that a formal power series in λ

$$(36.2) \qquad \varphi(s) = \sum_{n=0}^{\infty} \lambda^n\, \varphi_n(s)$$

satisfies (36.1). Then, substituting this in (36.1), we have

$$f(s) = \sum_{n=0}^{\infty} \lambda^n\, \varphi_n(s) - \lambda \int_a^s K(s, t) \sum_{n=0}^{\infty} \lambda^n\, \varphi_n(t)\, dt$$

$$= \varphi_0(s) + \sum_{n=1}^{\infty} \lambda^n \left[\varphi_n(s) - \int_a^s K(s, t)\, \varphi_{n-1}(t)\, dt \right]$$

hence we obtain, by a comparison of coefficients, the following relations

$$(36.3) \qquad \begin{aligned} &\varphi_0(s) = f(s)\,, \\ &\varphi_n(s) = \int_a^s K(s, t)\, \varphi_{n-1}(t)\, dt \qquad (n \geq 1) \end{aligned}$$

Next we shall prove that the power series (36.2) with the coefficients (36.3) is a genuine solution of (36.1). Set

$$(36.4) \qquad \sup_s |f(s)| = M\,, \qquad \sup_{(s,t)} |K(s, t)| = N$$

Then we have

$$|\varphi_0(s)| \leq M\,, \qquad |\varphi_1(s)| \leq MN(s - a)$$

145

and hence

$$|\varphi_2(s)| \leq MN \int_a^s |K(s, t)| (t - a)\, dt \leq MN^2 \frac{(s - a)^2}{2}$$

Proceeding in this manner we prove by mathematical induction that

$$(36.5) \qquad |\varphi_n(s)| \leq MN^n \frac{(s - a)^n}{n!} \qquad (n = 1, 2, 3, \cdots)$$

In fact, from the assumption

$$|\varphi_{n-1}(s)| \leq MN^{n-1} \frac{(s - a)^{n-1}}{(n - 1)!}$$

it follows immediately that

$$|\varphi_n(s)| \leq \frac{MN^{n-1}}{(n - 1)!} \int_a^s |K(s, t)| (t - a)^{n-1}\, dt$$

$$= MN^n \frac{(s - a)^n}{n!}$$

The equation (36.5) implies that, for $|\lambda| < \infty$, the series (36.2) converges absolutely and uniformly on the interval $a \leq s \leq b$. Hence the formal solution (36.2) is a genuine solution of (36.1).

Uniqueness of solutions. Suppose that there exist two solutions of the equation (36.1). Then the difference $\psi(s)$ satisfies the equation

$$(36.6) \qquad \psi(s) = \lambda \int_a^s K(s, t)\, \psi(t)\, dt$$

Set

$$\sup_s |\psi(s)| = L$$

Then (36.6) implies

$$|\psi(s)| \leq |\lambda| L \int_a^s |K(s, t)|\, dt \leq |\lambda| LN(s - a)$$

Substituting this in the right side of (36.6), we obtain

$$|\psi(s)| \leq |\lambda|^2 LN \int_a^s |K(s, t)| (t - a)\, dt$$

$$\leq L \frac{\{|\lambda| N(s - a)\}^2}{2!}$$

Repeating the same procedure, we obtain

$$|\psi(s)| \le L \frac{\{|\lambda|\, N(s-a)\}^n}{n!}\,, \qquad n = 1, 2, \cdots$$

Letting $n \to \infty$, we obtain that $\psi(s) \equiv 0$.

37. *Resolvent kernels*

We shall give another expression of the solution $\varphi(s)$ as follows. From (36.3) it follows that

$$\varphi_1(s) = \int_a^s K(s, t)\, f(t)\, dt$$

$$\varphi_2(s) = \int_a^s K(s, t)\, \varphi_1(t)\, dt$$

and hence

$$\varphi_2(s) = \int_a^s dr \int_a^r K(s, r)\, K(r, t)\, f(t)\, dt$$

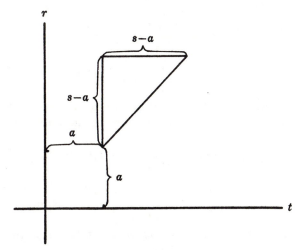

Fig. 2.

On the other hand, for $F(s, r, t) = K(s, r)\, K(r, t)$, the *Dirichlet integral formula* holds

(37.1) $$\int_a^s dr \int_a^r F(s, r, t)\, dt = \int_a^s dt \int_t^s F(s, r, t)\, dr$$

since the integration is taken over the triangular set in Fig. 2.

Accordingly, by means of the iterated kernel

(37.2) $$K^{(2)}(s, t) = \int_t^s K(s, r)\, K(r, t)\, dr$$

we can write

(37.3) $$\varphi_2(s) = \int_a^s K^{(2)}(s, t)\, f(t)\, dt$$

Repeating the same argument, we obtain the *iterated kernels*

(37.4)
$$K^{(1)}(s, t) = K(s, t)$$
$$K^{(n)}(s, t) = \int_t^s K(s, r)\, K^{(n-1)}(r, t)\, dr \qquad (n = 2, 3, \cdots)$$

and we can write $\varphi_n(s)$ as follows

(37.5) $$\varphi_n(s) = \int_a^s K^{(n)}(s, t)\, f(t)\, dt$$

Accordingly, setting

(37.6) $$\Gamma(s, t; \lambda) = K(s, t) + \lambda K^{(2)}(s, t) + \cdots + \lambda^{n-1} K^{(n)}(s, t) + \cdots$$

we obtain that

(37.7) $$\varphi(s) = f(s) + \lambda \int_a^s \Gamma(s, t; \lambda)\, f(t)\, dt$$

which means that, the solution $\varphi(s)$ of the equation (36.1) is given by (37.7). It is an immediate consequence of the estimates

(37.8) $$|K^{(n)}(s, t)| \leqq N^{n-1} \frac{1}{(n-1)!} \int_t^s N|s - t|^{n-1}\, dt \leqq N^n \frac{|s - t|^n}{n!},$$

which can be proved in the same way as (36.5) by mathematical induction, that the series (37.6) converges uniformly with respect to (s, t) for $|\lambda| < \infty$.

The function $\Gamma(s, t; \lambda)$ is called the *resolvent kernel* of the kernel $K(s, t)$. Obviously, $\Gamma(s, t; \lambda)$ satisfies

(37.9)
$$\Gamma(s, t; \lambda) = K(s, t) + \lambda \int_t^s K(s, r)\, \Gamma(r, t; \lambda)\, dr$$
$$= K(s, t) + \lambda \int_t^s \Gamma(s, r; \lambda)\, K(r, t)\, dr$$

since, together with (37.4), we have the relation

$$(37.4') \qquad K^{(n)}(s, t) = \int_t^s K^{(n-1)}(s, r)\, K(r, t)\, dr \qquad (n \geqq 2)$$

We can prove in the same way as we have proved the uniqueness of the solutions of (36.1) that for given $K(s, t)$ the solution $\Gamma(s, t; \lambda)$ of the equation (37.9) is uniquely determined. From this follows the fact that the resolvent kernel $\Gamma(s, t; \lambda)$ of the kernel $K(s, t)$ is also defined by (37.9). Similarly, we can prove that for given $\Gamma(s, t; \lambda)$ the equation (37.9) admits a unique solution $K(s, t)$, in other words, the resolvent kernel of $\Gamma(s, t; \lambda)$ is $K(s, t)$ itself.

Hence, we can conclude that if $\Gamma(s, t; \lambda)$ is the resolvent kernel of $K(s, t)$, then the resolvent kernel of $\Gamma(s, t; \lambda)$ is the kernel $K(s, t)$ itself, which is known as the *reciprocity theorem*.

EXAMPLE.

$$s = \varphi(s) + \int_0^s (s - t)\, \varphi(t)\, dt \qquad (0 \leqq s \leqq 1)$$

The iterated kernels are as follows:

$$
\begin{aligned}
K(s, t) \quad &= s - t \\
K^{(2)}(s, t) &= \int_s^t (s - r)\,(r - t)\, dr = \frac{(s - t)^3}{2 \cdot 3} \\
K^{(3)}(s, t) &= \frac{1}{2 \cdot 3} \int_s^t (s - r)^3 (r - t)\, dr = \frac{(s - t)^5}{5!}
\end{aligned}
$$

$$\dotfill$$

$$K^{(n)}(s, t) = \frac{(s - t)^{2n-1}}{(2n - 1)!}$$

$$\dotfill$$

Hence the solution $\varphi(s)$ is given by

$$
\begin{aligned}
\varphi(s) &= s - \int_0^s \left\{ (s - t) - \frac{(s - t)^3}{3!} + \frac{(s - t)^5}{5!} - \cdots \right\} t\, dt \\
&= s + \left[\frac{(s - t)^3}{3!} - \frac{(s - t)^5}{5!} + \frac{(s - t)^7}{7!} - \cdots \right]_0^s = \sin s
\end{aligned}
$$

38. *Application to linear differential equations*

We consider a linear differential equation of nth order, in an unknown y,

$$(38.1) \qquad \frac{d^n y}{dx^n} + p_1(x) \frac{d^{n-1} y}{dx^{n-1}} + \cdots + p_n(x) y = f(x)$$

where $f(x)$ and the coefficients $p_h(x)$ are assumed to be continuous in a neighbourhood of the point $x = 0$.

Setting

$$(38.2) \qquad \frac{d^n y}{dx^n} = z(x)$$

we obtain that

$$\frac{d^{n-1}y}{dx^{n-1}} = \int_0^x z \, dx + C_1$$

$$\dots \dots \dots \dots \dots \dots$$

$$y = \int_0^x \cdots \int_0^x z \, dx^n + C_1 \frac{x^{n-1}}{(n-1)!} + C_2 \frac{x^{n-2}}{(n-2)!} + \cdots + C_n$$

Accordingly the equation (38.1) is reduced to an equation, in unknown z, of the form

$$z + p_1(x) \int_0^x z \, dx + \cdots + p_n(x) \int_0^x \cdots \int_0^x z \, dx^n$$

$$(38.3) \qquad = f(x) + \sum_{h=1}^n c_h f_h(x)$$

$$f_h(x) = p_h(x) + \frac{x}{1!} p_{h+1}(x) + \cdots + \frac{x^{n-h}}{(n-h)!} p_n(x)$$

On the other hand, for any continuous function $\varphi(s)$, there holds the relation

$$(38.4) \qquad \int_0^x \cdots \int_0^x \varphi(s) \, ds^n = \int_0^x \frac{(x-s)^{n-1}}{(n-1)!} \varphi(s) \, ds$$

as is easily proved by differentiation. Accordingly the equation (38.3) and therewith the equation (38.1) is equivalent to an equation of the form

$$(38.5) \qquad z(x) + \int_0^x \left[p_1(x) + p_2(x)(x-s) + \cdots + p_n(x) \frac{(x-s)^{n-1}}{(n-1)!} \right] z(s) \, ds$$

$$= f(x) + \sum_{h=1}^n c_h f_h(x)$$

The uniqueness of the solution $y(x)$ of the initial value problem for the nth order linear differential equation (38.1) corresponds exactly to the uniqueness of the solution $z(x)$ of the Volterra integral equation (38.5). The constants $c_1, c_2, \cdots, c_n$ in (38.5) are determined by the initial conditions for (38.1).

REMARK. The above argument suggests that we can define a linear differential equation of infinite order by

$$(38.6) \quad z(x) + \int_0^x \left[p_1(x) + p_2(x)(x-s) + \cdots \right. $$
$$\left. + p_n(x) \frac{(x-s)^{n-1}}{(n-1)!} + \cdots \right] z(s) \, ds$$
$$= f(x) + \sum_{h=1}^{\infty} c_h f_h(x)$$

where

$$(38.7) \qquad f_h(x) = p_h(x) + \frac{x}{1!} p_{h+1}(x) + \cdots + \frac{x^{n-h}}{(n-h)!} p_n(x)$$

If $\sup_x |p_h(x)| \leqq A$ $(h = 1, 2, 3, \cdots)$ in an interval $[0, a]$, then $\sup_x |f_h(x)| \leqq A e^x$. Accordingly, if moreover $\sum_{h=1}^{\infty} |c_h| < \infty$, then the series on the right side of (38.6) converges absolutely and uniformly on the interval $[0, a]$ as well as the kernel in the bracket. Thus we can find a unique solution $z(x)$ of the equation (38.6) in the same way as in Part 37.

39. *The singular kernel $P(s, t)/(s-t)^\alpha$*

Let $P(s, t)$ be a continuous function defined on a domain $a \leqq s \leqq b$, $a \leqq t \leqq b$. We consider a kernel of the form

$$(39.1) \qquad\qquad K(s, t) = P(s, t)/(s-t)^\alpha$$

where $0 < \alpha < 1$. We have

$$K^{(2)}(s, t) = \int_t^s \frac{P(s, r) P(r, t)}{(s-r)^\alpha (r-t)^\alpha} \, dr$$

If we make the transformation of variable

$$r = t + (s-t) w$$

we obtain

$$K^{(2)}(s, t) = (s-t)^{1-2\alpha} \int_0^1 \frac{P(s, t+(s-t)w) P(t+(s-t)w, t)}{(1-w)^\alpha w^\alpha} \, dw$$

Hence, $K^{(2)}(s, t)$ may be written as

$$K^{(2)}(s, t) = \frac{Q^{(2)}(s, t)}{(s-t)^{2\alpha-1}}$$

where $Q^{(2)}(s, t)$ is a continuous function. By the same argument, we can prove, successively, that the iterated kernel $K^{(n)}(s, t)$ with $K^{(1)}(s, t) = K(s, t)$ is of the form

$$K^{(n)}(s, t) = \frac{Q^{(n)}(s, t)}{(s - t)^{n\alpha - (n-1)}} \qquad (n = 1, 2, 3, \cdots)$$

where $Q^{(n)}(s, t)$ is a continuous function and $Q^{(1)}(s, t) = P(s, t)$. Since $0 < \alpha < 1$, there exists a positive integer n such that $n\alpha - (n-1) < 0$; hence, for $m \geq n$, $K^{(m)}(s, t)$ are all continuous. Accordingly, if we discard its first n terms, then, for $|\lambda| < \infty$, the resolvent kernel (37.6) converges absolutely and uniformly with respect to (s, t).

Let

$$\varphi_1(s) = \int_a^s K(s, t) f(t)\, dt$$

$$\varphi_2(s) = \int_a^s K^{(2)}(s, t) f(t)\, dt$$

$$\cdots\cdots\cdots\cdots\cdots$$

and put

$$\sup_s |f(s)| = M$$

Then we obtain that, for $m \leq n$,

$$|\varphi_m(s)| \leq C_m M \left| \int_a^s (s - t)^{m-1-m\alpha}\, dt \right|$$

$$\leq C_m M \frac{|s - a|^{m-m\alpha}}{m - m\alpha}$$

where $C_m = \sup_{(s, t)} |Q^{(m)}(s, t)|$. Thus, we see that, for $|\lambda| < \infty$,

$$\varphi(s) = f(s) + \lambda \int_a^s \Gamma(s, t; \lambda) f(t)\, dt$$

converges absolutely and uniformly with respect to s. Hence $\varphi(s)$ is a continuous solution of the equation (36.1) with the singular kernel (39.1).

We shall next prove the uniqueness of the solution. To prove this, suppose that there exist two solutions. Then the difference $\psi(s)$ satisfies

$$\psi(s) = \lambda \int_a^s K(s, t)\, \psi(t)\, dt$$

and hence, satisfies

$$\psi(s) = \lambda^n \int_a^s K^{(n)}(s, t)\, \psi(t)\, dt$$

Therefore, as in Part 36, we can prove that $\psi(s) \equiv 0$, because of the continuity of $K^{(n)}(s, t)$. In the above, the necessary relation

$$\int_a^s K^{(n-1)}(s, r)\, dr \int_a^r K(r, t)\, \psi(t)\, dt = \int_a^s \left\{ \int_t^s K^{(n-1)}(s, r)\, K(r, t)\, dr \right\} \psi(t)\, dt$$

can be proved in the same manner as in Part 37.

§2. Volterra integral equations of the first kind

40. *Reduction to integral equations of the second kind*

Let $K(s, t)$ be a complex-valued continuous function defined on a domain $a \leqq s \leqq b$, $a \leqq t \leqq b$, and have the continuous partial derivative $K_s(s, t)$. Let $f(s)$ be a complex-valued continuously differentiable function defined on the interval $a \leqq s \leqq b$. We consider an equation, in unknown $\varphi(s)$, of the form

$$(40.1) \qquad f(s) = \int_a^s K(s, t)\, \varphi(t)\, dt$$

Such an equation is called a *Volterra integral equation of the first kind*. We shall be concerned with the reduction of such an equation to that of the second kind.

Let $\varphi(s)$ be a continuous solution of the equation (40.1). Then $\varphi(s)$ satisfies the equation

$$(40.2) \qquad f'(s) = K(s, s)\, \varphi(s) + \int_a^s K_s(s, t)\, \varphi(t)\, dt$$

If $K(s, s) \neq 0$ in the interval $a \leqq s \leqq b$, then the equation (40.2) is equivalent to an equation of the second kind

$$(40.2') \qquad \frac{f'(s)}{K(s, s)} = \varphi(s) + \int_a^s \frac{K_s(s, t)}{K(s, s)}\, \varphi(t)\, dt$$

Hence the equation (40.2) admits a unique continuous solution, as was shown in Part 36. Moreover, if we assume further that $f(a) = 0$, then the solution of (40.2) also satisfies the equation (40.1); for then the equation (40.1) is derived from (40.2) by integration. This result is summarized in the following theorem.

THEOREM 40.1. Let $f(a)=0$ and $K(a, a) \neq 0$. Then the equation (40.1) admits a unique continuous solution in the interval $[a, b']$, $a < b' \leqq b$, if $f'(s)$ and $K_s(s, t)$ are continuous, and $K(s, s) \neq 0$ for $a \leqq s \leqq b'$, $a \leqq t \leqq b'$. The solution of (40.1) is given by the equation (40.2').

If $K(a, a) = 0$, the above argument can not be used. However, if $K(s, s) \equiv 0$, we can derive from (40.2), by differentiation, an equation

$$(40.3) \qquad f''(s) = K_s(s, s) \varphi(s) + \int_a^s K_{ss}(s, t) \varphi(t) \, dt$$

provided that $f''(s)$ and $K_{ss}(s, t)$ exist and are continuous. Hence, if $f'(a) = 0$, the equation (40.2) is equivalent to (40.3). On the other hand, (40.2) is equivalent to the equation (40.1) under the assumption $f(a) = 0$. Therefore, the equation (40.1) can be reduced to an equation of the second kind

$$(40.2'') \qquad \frac{f''(s)}{K_s(s, s)} = \varphi(s) + \int_a^s \frac{K_{ss}(s, t)}{K_s(s, s)} \varphi(t) \, dt$$

in the interval $[a, b']$, $a < b' \leqq b$, in which $K_s(s, s)$ does not vanish. Accordingly, if $f(a) = f'(a) = 0$, then, as in the previous case, the equation (40.1) admits a unique continuous solution $\varphi(s)$.

In the case when $K_s(s, s) \equiv 0$, the above procedure is also valid, provided that $f(s)$ and $K(s, t)$ are sufficiently many times differentiable. However, for the case when $K(a, a) = 0$, and $K(s, s) \not\equiv 0$, our procedure is no more applicable, and a more delicate treatment is needed. We will not enter into the details and refer the reader to T. Lalésco, *Introduction à la théorie des équations intégrales*, Paris, 1912, p. 103.

41. *Abel integral equations*

Let $G(s, t)$ be a complex-valued continuous function defined on a domain $a \leqq s \leqq b$, $a \leqq t \leqq b$, and

$$(41.1) \qquad G(s, s) \neq 0$$

on this domain. We consider a Volterra integral equation of the first kind with the singular kernel

$$(41.2) \qquad K(s, t) = G(s, t)/(s - t)^\alpha \qquad (0 < \alpha < 1)$$

namely,

$$(41.3) \qquad \int_a^s \frac{G(s, t)}{(s - t)^\alpha} \varphi(t) \, dt = f(s)$$

Such an equation is called a *generalized Abel integral equation*. To solve the equation (41.3), multiply both sides by $1/(x - s)^{1-\alpha}$ and integrate from a to x. Then we have

$$\int_a^x \frac{ds}{(x - s)^{1-\alpha}} \left\{ \int_a^s \frac{G(s, t)\, \varphi(t)}{(s - t)^{\alpha}}\, dt \right\} = \int_a^x \frac{f(s)}{(x - s)^{1-\alpha}}\, ds$$

By making use of the Dirichlet integral formula (Part 37), we obtain

$$(41.4) \qquad \int_a^x \varphi(t)\, dt \left\{ \int_t^x \frac{G(s, t)}{(x - s)^{1-\alpha}(s - t)^{\alpha}}\, ds \right\} = \int_a^x \frac{f(s)}{(x - s)^{1-\alpha}}\, ds$$

Put

$$(41.5) \qquad \int_t^x \frac{G(s, t)}{(x - s)^{1-\alpha}(s - t)^{\alpha}}\, ds = K_1(x, t)$$

$$= \int_0^1 \frac{G(t + r(x - t), t)}{(1 - r)^{1-\alpha} r^{\alpha}}\, dr$$

and

$$(41.6) \qquad \int_a^x \frac{f(s)}{(x - s)^{1-\alpha}}\, ds = F(x)$$

$$= \int_0^1 (x - a)^{\alpha}\, \frac{f(a + r(x - a))}{(1 - r)^{1-\alpha}}\, dr$$

Then the equation (41.4) becomes

$$(41.4') \qquad \int_a^x K_1(x, t)\, \varphi(t)\, dt = F(x)$$

Accordingly, if $G_s(s, t)$ and $f'(s)$ are continuous, then, since $G(s, s) \neq 0$ by (41.1), the equation (41.4'), and hence (41.4), can be reduced to an equation of the second kind, as was shown in Part 40.

Next we shall prove that the unique continuous solution $\varphi(t)$ of the equation of the second kind is also a solution of (41.3). To prove this, let us put

$$h(s) = \int_a^s \frac{G(s, t)}{(s - t)^{\alpha}}\, \varphi(t)\, dt - f(s)$$

Then, it is sufficient to prove that

$$(41.7) \qquad \int_a^x \frac{h(s)}{(x - s)^{1-\alpha}}\, ds \equiv 0$$

implies

$$h(s) \equiv 0$$

In fact, multiplying both sides of (41.7) by $1/(y - x)^{\alpha}$ and integrating from a to y, we obtain, by the Dirichlet integral formula, that

$$\int_a^y h(s)\, ds \left\{ \int_0^1 \frac{1}{(1-r)^\alpha r^{1-\alpha}}\, dr \right\} \equiv 0$$

Since

(41.8)
$$\int_0^1 \frac{1}{(1-r)^\alpha r^{1-\alpha}}\, dr = \frac{\pi}{\sin \alpha \pi}$$

we obtain

$$\int_a^y h(s)\, ds \equiv 0$$

which implies

$$h(s) \equiv 0$$

The so-called Abel integral equation is the equation

(41.9)
$$\int_a^s \frac{\varphi(t)}{(s-t)^\alpha}\, dt = f(s)$$

which is the equation (41.3) with $G(s, t) \equiv 1$. In this case, $K_1(s, t) = \pi/\sin \alpha \pi$ and hence the equation can be reduced to

(41.10)
$$\frac{\pi}{\sin \alpha \pi} \int_a^x \varphi(t)\, dt = \int_a^x \frac{f(s)}{(x-s)^{1-\alpha}}\, ds$$

A solution $\varphi(x)$ of the equation (41.10) is easily obtained by differentiation as follows

(41.11)
$$\varphi(x) = \frac{\sin \alpha \pi}{\pi} \frac{\partial}{\partial x} \left\{ \int_a^x \frac{f(s)}{(x-s)^{1-\alpha}}\, ds \right\}$$

$$= \frac{\sin \alpha \pi}{\pi} \frac{\partial}{\partial x} \left\{ \int_0^1 (x-a)^\alpha \frac{f(a + r(x-a))}{(1-r)^{1-\alpha}}\, dr \right\}$$

$$= \frac{\sin \alpha \pi}{\pi} \left\{ \alpha(x-a)^{\alpha-1} \int_0^1 \frac{f(a + r(x-a))}{(1-r)^{1-\alpha}}\, dr \right.$$

$$\left. + (x-a)^\alpha \int_0^1 \frac{f'(a + r(x-a))}{(1-r)^{1-\alpha}}\, r\, dr \right\}$$

REMARK. Abel's integral equation was derived from the following problem in Mechanics: Let a material point move under the influence of the gravity along a smooth curve in a vertical plane. Let the time t which is required for the point to move along the curve from the vertical height x to a fixed point O on the curve be a given function $f(x)$. What is the equation of the curve?

Take the point O as the origin, the vertical line as x-axis and the horizontal line as y-axis. Let the equation of the curve be

$$y = F(x) \qquad [F(0) = 0]$$

Let the point move from a point with the height x to another point with the height $x - t$. Then its velocity is given by

$$\sqrt{(2g(x - t))}$$

where g is the acceleration due to the gravity. Since the infinitesimal length ds of the curve at the point is

$$ds = \sqrt{(1 + F'(t)^2)} \cdot dt$$

$F(t)$ must satisfy the integral equation

$$f(x) = \int_0^x \frac{\sqrt{(1 + F'(t)^2)}}{\sqrt{(2g(x - t))}} \, dt$$

provided that $F'(t)$ is continuous.

This equation can be solved by the procedure mentioned above. Hence we can obtain $\sqrt{(1 + F'(t)^2)}$, and therewith $F'(t)$. Furthermore using the condition $F(0) = 0$, we obtain $F(x)$.

CHAPTER 5

THE GENERAL EXPANSION THEOREM (WEYL-STONE-TITCHMARSH-KODAIRA'S THEOREM)

We shall be concerned with the singular boundary value problem for the equation

$$\frac{d}{dx}\left(p\frac{dz}{dx}\right) - rz + \lambda\rho z = 0$$

By means of the Liouville transformation

$$y = \sqrt[4]{(p\rho)}z \,, \qquad t = \int^x \sqrt{(\rho/p)}\,dx$$

this equation is changed into the form

$$\frac{d^2y}{dt^2} - q(t)y + \lambda y = 0$$

as was shown in Part 25. Accordingly, we shall treat the latter equation.

Let $q(x)$ be a real-valued continuous function in a finite or infinite open interval (a, b). We make no assumption concerning the behaviour of $q(x)$ at the boundary points; as $x \to a$ or $x \to b$, $q(x)$ may tend to finite limits, may tend to $\pm\infty$, or may have no limits. Thus we consider the general singular case,

$$y'' + \{\lambda - q(x)\}y = 0 \,, \qquad a < x < b$$

where λ is a complex parameter and both the boundary points a and b are, in general, singular. Let $y_1(x, \lambda)$, $y_2(x, \lambda)$ be a fundamental system of solutions determined by initial conditions at a point c, $a < c < b$,

$$y_1(c, \lambda) = 1 \,, \qquad y_1'(c, \lambda) = 0$$
$$y_2(c, \lambda) = 0 \,, \qquad y_2'(c, \lambda) = 1$$

Then, for "appropriate boundary conditions at $x = a$, $x = b$," there exists a "density matrix"

$$\begin{pmatrix} \rho_{11}(u) & \rho_{12}(u) \\ \rho_{21}(u) & \rho_{22}(u) \end{pmatrix} \qquad (-\infty < u < \infty)$$

such that every real-valued continuous function $f(x)$ in (a, b) with

$\int_a^b f(x)^2 dx < \infty$ can be expanded, in a sense which will be explained later, as follows

$$f(x) = \int_{-\infty}^{\infty} du \left\{ \sum_{j,k=1}^{2} \int_0^u y_j(x, u) \, d\rho_{jk}(u) \int_a^b f(s)y_k(s, u)ds \right\}$$

This is the *Weyl-Stone expansion theorem* which has been completed by Titchmarsh and Kodaira by giving an explicit formula for the density matrix. The general expansion theorem thus obtained enables us to give a unified treatment of the classical expansions in terms of special functions, such as the Fourier series expansion, the Fourier integral, the Hermite polynomials expansion, the Laguerre polynomials expansion and the Bessel functions expansion.

We shall prove the general expansion theorem as a limiting case of the Hilbert-Schmidt expansion theorem. For the sake of simplicity, we consider the case when $a = -\infty$, and $b = \infty$. The case when (a, b) is a finite open interval, or $a \neq -\infty$ or $b \neq \infty$ may be discussed similarly.

§ 1. Classification of singular boundary points

42.   *Green's formula*

We consider the differential equation

$$(42.1) \qquad y'' + \{\lambda - q(x)\}y = 0 \qquad (-\infty < x < \infty)$$

where $q(x)$ is real-valued and continuous in the open infinite interval $(-\infty, \infty)$ and λ is a complex parameter. Let $y_1(x, \lambda)$, $y_2(x, \lambda)$ be the fundamental system of solutions of (42.1) determined by the initial conditions

$$(42.2) \qquad \begin{aligned} y_1(0, \lambda) &= 1, & y_1'(0, \lambda) &= 0 \\ y_2(0, \lambda) &= 0, & y_2'(0, \lambda) &= 1 \end{aligned}$$

Then, as was shown in Part 5, the functions $y_1(x, \lambda)$ and $y_2(x, \lambda)$, together with their derivatives $y_1'(x, \lambda)$ and $y_2'(x, \lambda)$, are regular in the complex domain $|\lambda| < \infty$ as functions of λ, in other words, they are entire functions of λ. Moreover,

$$(42.3) \qquad y_1(x, \bar{\lambda}) \equiv \overline{y_1(x, \lambda)}, \qquad y_2(x, \bar{\lambda}) \equiv \overline{y_2(x, \lambda)}$$

which will be proved as follows. Taking the complex conjugate of

$$y_1'' + \{\lambda - q(x)\}y_1 = 0$$

we obtain

$$\bar{y}_1'' + \{\bar{\lambda} - q(x)\}\bar{y}_1 = 0$$

Hence $\overline{y_1(x, \lambda)}$ is a solution of the differential equation

$$z'' + \{\bar{\lambda} - q(x)\}z = 0$$

satisfying the initial conditions

$$z(0, \bar{\lambda}) = 1 , \qquad z'(0, \bar{\lambda}) = 0$$

On the other hand, according to its definition, $z = y_1(x, \bar{\lambda})$ is also a solution of the above equation satisfying the same initial conditions. Thus, the uniqueness of the solution (Part 6) implies

$$y_1(x, \bar{\lambda}) \equiv \overline{y_1(x, \lambda)}$$

The corresponding property for y_2 is proved in the same manner.

For the sake of simplicity, we write

(42.4) $$L_x = q(x) - d^2/dx^2$$

Let $F(x, \lambda)$ be a solution of the equation (42.1) and $G(x, \lambda')$ a solution of the equation with a parameter λ' instead of λ, that is,

(42.5) $$L_x F(x, \lambda) = \lambda F(x, \lambda) , \qquad L_x G(x, \lambda') = \lambda' G(x, \lambda')$$

for $0 \leqq x < b < \infty$. Then we have Green's formula

(42.6) $$(\lambda' - \lambda) \int_0^x FG\,dx = \int_0^x \{FL_x G - GL_x F\}\,dx$$
$$= \int_0^x \{-FG'' + F''G\}\,dx = W_0(F, G) - W_x(F, G)$$

where $W_x(H, K)$ is the Wronskian of $H(x)$ and $K(x)$,

(42.7) $$W_x(H, K) = H(x)K'(x) - H'(x)K(x)$$

Setting $\lambda = \lambda'$ in (42.6), we see that $W_x[F(x, \lambda), G(x, \lambda)]$ does not depend on x. Hence we may write

(42.8) $$W_x[F(x, \lambda), G(x, \lambda)] = \omega(\lambda)$$

In particular, let

(42.9) $$F(x, \lambda) \equiv y_2(x, \lambda)$$

and let $G(x, \lambda)$ be a solution of $L_x G = \lambda G$ satisfying the initial conditions at $x = b$,

(42.10) $$G(b, \lambda) = -\sin \beta , \qquad G'(b, \lambda) = \cos \beta$$

where β is a real number, independent of either b or λ. Then we have

$$(42.11) \qquad \omega(\lambda) = W_x[F(x, \lambda), G(x, \lambda)] = W_b[F(x, \lambda), G(x, \lambda)]$$
$$= y_2(b, \lambda) \cos \beta + y_2'(b, \lambda) \sin \beta$$

Therefore, $\omega(\lambda_0) = 0$ if and only if λ_0 is an eigenvalue, with the eigenfunction $y_2(x, \lambda_0)$, of the boundary value problem

$$\begin{cases} L_x y = \lambda y \\ 1 \cdot y(0) + 0 \cdot y'(0) = 0 \, , \quad y(b) \cos \beta + y'(b) \sin \beta = 0 \end{cases}$$

As was shown in Part 21 (Theorem 21.1), every eigenvalue λ_0 of this boundary value problem is real. This proves the following

THEOREM 42.1. Let b be a positive number and β a real number. Then every zero of the entire function

$$(42.12) \qquad y_2(b, \lambda) \cos \beta + y_2'(b, \lambda) \sin \beta$$

is real. The same is true for $y_1(b, \lambda) \cos \beta + y_1'(b, \lambda) \sin \beta$.

43. Limit point case and limit circle case

For any number e, the expression $y_1(x, \lambda) + e y_2(x, \lambda)$ satisfies the equation (42.1). We now choose e so that $y_1(x, \lambda) + e y_2(x, \lambda)$ satisfies the boundary condition

$$(43.1) \quad \{y_1(b, \lambda) + e y_2(b, \lambda)\} \cos \beta + \{y_1'(b, \lambda) + e y_2'(b, \lambda)\} \sin \beta = 0$$

at the point b and denote it by $l_b(\lambda)$. Then $l_b(\lambda)$ must satisfy

$$(43.2) \qquad l_b(\lambda) = - \frac{y_1(b, \lambda) \cos \beta + y_1'(b, \lambda) \sin \beta}{y_2(b, \lambda) \cos \beta + y_2'(b, \lambda) \sin \beta}$$

Since $y_1(b, \lambda)$, $y_2(b, \lambda)$, $y_1'(b, \lambda)$ and $y_2'(b, \lambda)$ are all entire functions of λ, $l_b(\lambda)$ is a meromorphic function of λ. Furthermore, by Theorem 42.1, all the poles of $l_b(\lambda)$ lie on the real-axis of the λ-plane. We write

$$(43.3) \qquad l_b(\lambda, z) = - \frac{y_1(b, \lambda)z + y_1'(b, \lambda)}{y_2(b, \lambda)z + y_2'(b, \lambda)}$$

If b and λ are fixed, and z varies, (43.3) may be written as

$$(43.3') \qquad l = \frac{\gamma z + \delta}{\varepsilon z + \eta}$$

where γ, δ, ε and η are fixed. Since

$$|\varepsilon \delta - \gamma \eta| = |y_1(b, \lambda)y_2'(b, \lambda) - y_1'(b, \lambda)y_2(b, \lambda)| \neq 0$$

the linear transformation (43.3′) is a one-to-one conformal mapping which transforms circles into circles; straight lines being considered as circles with infinite radii. Therefore, if $\mathscr{I}(\lambda)=v\neq0$, then $l_b(\lambda,z)$ varies on a circle $C_b(\lambda)$, with a finite radius, in the l-plane, as z varies over the real-axis of the z-plane.

The centre and the radius of the circle $C_b(\lambda)$ will be determined as follows. The centre of the circle is the symmetric point of the point at infinity with respect to the circle. Thus if we set

$$l_b(\lambda, z') = \infty$$

$$l_b(\lambda, z'') = \text{the centre of } C_b(\lambda)$$

z'' must be the symmetric point of z' with respect to the real-axis of the z-plane, namely, $z' = \bar{z}''$. On the other hand,

$$(43.4) \qquad l_b\!\left(\lambda, -\frac{y_2'(b, \lambda)}{y_2(b, \lambda)}\right) = \infty$$

Therefore, the centre of the circle $C_b(\lambda)$ is given by

$$(43.5) \qquad l_b\!\left(\lambda, -\frac{\overline{y_2'(b, \lambda)}}{y_2(b, \lambda)}\right) = -\frac{W_b(y_1, \bar{y}_2)}{W_b(y_2, \bar{y}_2)}$$

The radius $r_b(\lambda)$ of the circle $C_b(\lambda)$ is equal to the distance between the centre of $C_b(\lambda)$ and the point $l_b(\lambda, 0)$ on the circle $C_b(\lambda)$. Hence

$$
\begin{aligned}
r_b(\lambda) &= \left| \frac{y_1'(b, \lambda)}{y_2'(b, \lambda)} - \frac{W_b(y_1, \bar{y}_2)}{W_b(y_2, \bar{y}_2)} \right| \\
&= \left| \frac{y_1'y_2\bar{y}_2' - y_1'\bar{y}_2 y_2' - y_2'y_1\bar{y}_2' + y_2'\bar{y}_2 y_1'}{y_2' W_b(y_2, \bar{y}_2)} \right| \\
&= \left| \frac{-\bar{y}_2'(y_1 y_2' - y_1' y_2)}{y_2' W_b(y_2, \bar{y}_2)} \right| = \left| \frac{W_b(y_1, y_2)}{W_b(y_2, \bar{y}_2)} \right|
\end{aligned}
$$

On the other hand, by virtue of (42.2),

$$W_b(y_1, y_2) = W_0(y_1, y_2) = 1$$

Further, by virtue of (42.2), (42.3) and by making use of (42.6), we have

$$(43.6) \qquad 2v\int_0^b |y_2(x, \lambda)|^2 dx = 2v\int_0^b y_2(x, \lambda)y_2(x, \bar{\lambda})dx$$

$$= iW_0[y_2(x, \lambda), y_2(x, \bar{\lambda})] - iW_b[y_2(x, \lambda), y_2(x, \bar{\lambda})] \ (i = \sqrt{-1})$$

$$= -iW_b[y_2(x, \lambda), y_2(x, \bar{\lambda})], \ v = \mathscr{I}(\lambda)$$

Therefore, we obtain

$$(43.7) \qquad r_b(\lambda) = \frac{1}{2|v| \displaystyle\int_0^b |y_2(x, \lambda)|^2 dx}, \qquad \mathscr{I}(\lambda) = v \neq 0$$

We shall now prove the following

THEOREM 43.1. If $v = \mathscr{I}(\lambda) > 0$, then the interior of the circle $C_b(\lambda)$ is mapped onto the lower half plane of the z-plane by the transformation (43.3).

Proof. Since the real axis of the z-plane is the image of the circle $C_b(\lambda)$ by the transformation (43.3), the interior of $C_b(\lambda)$ is mapped onto either the upper half plane or the lower half plane of the z-plane, and further, the point at infinity of the l-plane is mapped onto the point $-y_2'(b, \lambda)/y_2(b, \lambda)$ of the z-plane.

On the other hand, by making use of (42.3) and (43.6), we obtain

$$(43.8) \qquad \mathscr{I}\left(-\frac{y_2'(b, \lambda)}{y_2(b, \lambda)}\right)$$

$$= \frac{i}{2}\left\{\frac{y_2'(b, \lambda)}{y_2(b, \lambda)} - \frac{\overline{y_2'(b, \lambda)}}{\overline{y_2(b, \lambda)}}\right\}$$

$$= \frac{-i}{2}\frac{W_b(y_2, \bar{y}_2)}{|y_2(b, \lambda)|^2} = \frac{v\displaystyle\int_0^b |y_2(x, \lambda)|^2 dx}{|y_2(b, \lambda)|^2} > 0$$

This means that $-y_2'(b, \lambda)/y_2(b, \lambda)$ belongs to the upper half plane of the z-plane. Hence the point at infinity, which is not contained in the interior of $C_b(\lambda)$, is mapped into the upper half plane. This proves the theorem, q.e.d.

Since $W_0(y_0, y_1) = 1$, the transformation (43.3) has a unique inverse which is given by

$$(43.9) \qquad z = -\frac{y_2'(b, \lambda)l_b + y_1'(b, \lambda)}{y_2(b, \lambda)l_b + y_1(b, \lambda)}$$

In view of Theorem 43.1, if $\mathscr{I}(\lambda) = v > 0$, l belongs to the interior of the circle $C_b(\lambda)$ if and only if $\mathscr{I}(z) < 0$, namely, $i(z - \bar{z}) > 0$. From (43.9) it follows that

$$i(z - \bar{z}) = i\left\{-\frac{y_2'(b, \lambda)l + y_1'(b, \lambda)}{y_2(b, \lambda)l + y_1(b, \lambda)} + \frac{\overline{y_2'(b, \lambda)}\bar{l} + \overline{y_1'(b, \lambda)}}{\overline{y_2(b, \lambda)}\bar{l} + \overline{y_1(b, \lambda)}}\right\}$$

$$= \frac{iW_b(y_1 + ly_2, \bar{y}_1 + \bar{l}\bar{y}_2)}{|y_2(b, \lambda)l + y_1(b, \lambda)|^2}$$

Therefore, $\mathscr{I}(z) < 0$ if and only if

$$i W_b(y_1 + l y_2, \bar{y}_1 + \bar{l}\bar{y}_2) > 0$$

By Green's formula (42.6), we have

$$2v \int_0^b |y_1 + l y_2|^2 dx$$
$$= i\{ W_0(y_1 + l y_2, \bar{y}_1 + \bar{l}\bar{y}_2) - W_b(y_1 + l y_2, \bar{y}_1 + \bar{l}\bar{y}_2)\}$$

We obtain further by (42.2)

$$W_0(y_1 + l y_2, \bar{y}_1 + \bar{l}\bar{y}_2)$$
$$= W_0(y_1, \bar{y}_1) + W_0(y_2, \bar{y}_1)l + W_0(y_1, \bar{y}_2)\bar{l} + W_0(y_2, \bar{y}_2)|l|^2$$
$$= - l + \bar{l} = - 2i\mathscr{I}(l)$$

Consequently, we obtain the following

THEOREM 43.2. If $v = \mathscr{I}(\lambda) > 0$, then l belongs to the interior of the circle $C_b(\lambda)$ if and only if

$$(43.10) \qquad \int_0^b |y_1(x, \lambda) + l y_2(x, \lambda)|^2 dx < \frac{\mathscr{I}(l)}{v}$$

and l lies on the circle $C_b(\lambda)$ if and only if

$$(43.10') \qquad \int_0^b |y_1(x, \lambda) + l y_2(x, \lambda)|^2 dx = \frac{\mathscr{I}(l)}{v}$$

REMARK. It is easy to see that Theorem 43.2 also holds when $v = \mathscr{I}(\lambda) < 0$.

If, in particular, l belongs to the interior of the circle $C_b(\lambda)$ and $0 < b' < b$, then

$$\int_0^{b'} |y_1 + l y_2|^2 dx \leqq \int_0^b |y_1 + l y_2|^2 dx < \frac{\mathscr{I}(l)}{v}$$

Hence, from Theorem 43.2, we have the following

THEOREM 43.3. If $v = \mathscr{I}(\lambda) \neq 0$, and $0 < b' < b$, then the set

$$\overline{C_b(\lambda)} \subseteq \overline{C_{b'}(\lambda)}$$

where $\overline{C_b(\lambda)}$ is the set composed of the circle $C_b(\lambda)$ and its interior.

This theorem implies that, if $v = \mathscr{I}(\lambda) \neq 0$, then the set

$$(43.11) \qquad \bigcap_{b>0} \overline{C_b(\lambda)} = C_\infty(\lambda)$$

is either a point or a closed circle with a non-zero finite radius. According as $C_\infty(\lambda)$ is a point or a circle, the singular boundary point $x = \infty$ is said to be in the *limit point case* or the *limit circle case*. According to this definition, the classification seems to depend upon both $q(x)$ and λ. However, it depends only on $q(x)$, as is shown in the following

Theorem 43.4. (i) If for some λ_0, $\mathscr{F}(\lambda_0) = v \neq 0$, the point $x = \infty$ is in the limit circle case, then, for every λ, every solution $y(x)$ of the equation

$$(43.12) \qquad y'' + \{\lambda - q(x)\}y = 0 , \qquad 0 \leq x < \infty$$

satisfies

$$(43.13) \qquad \int_0^\infty |y(x, \lambda)|^2 dx < \infty$$

(ii) If for some λ_0, $\mathscr{F}(\lambda_0) = v \neq 0$, every solution of the equation

$$(43.14) \qquad y'' + \{\lambda_0 - q(x)\}y = 0 , \qquad 0 \leq x < \infty$$

satisfies (43.13) with $\lambda = \lambda_0$, then the point $x = \infty$ is in the limit circle case for this λ_0.

Remark 1. According to Theorem 43.4, the classification is independent of λ. Thus, the point $x = \infty$ is in the limit circle case if and only if, for every λ, every solution $y(x)$ of (43.12) satisfies

$$\int_0^\infty |y(x)|^2 dx < \infty$$

The point $x = \infty$ is in the limit point case if and only if, for every λ, there exists at least one solution of (43.12) such that

$$\int_0^\infty |y(x)|^2 dx = \infty$$

Remark 2. Even if the point $x = \infty$ is in the limit point case, there exists, for every λ, $\mathscr{F}(\lambda) = v \neq 0$, at least one solution of (43.12) such that

$$\int_0^\infty |y(x)|^2 dx < \infty$$

In fact, from Theorem 43.2, it follows that

$$\int_0^\infty |y_1(x, \lambda) + ly_2(x, \lambda)|^2 dx \leq \frac{\mathscr{F}(l)}{v} < \infty$$

where $l = C_\infty(\lambda)$.

Proof of (ii). By the assumption, we have

$$\int_0^\infty |y_2(x, \lambda_0)|^2 dx < \infty$$

Hence, by virtue of (43.7), the radius $r_b(\lambda_0)$ of the circle $C_b(\lambda_0)$ remains positive as $b \to \infty$. Thus the proof is completed.

Proof of (i). We shall first prove that every solution $z(x)$ of (43.14) satisfies

$$\int_0^\infty |z(x)|^2 dx < \infty$$

To prove this, let l' and l'' be any two distinct points in $C_\infty(\lambda_0)$. Then, in view of Theorem 43.2,

$$\int_0^\infty |y_1(x, \lambda_0) + l' y_2(x, \lambda_0)|^2 dx \leqq \frac{\mathscr{F}(l')}{v} < \infty$$

$$\int_0^\infty |y_1(x, \lambda_0) + l'' y_2(x, \lambda_0)|^2 dx \leqq \frac{\mathscr{F}(l'')}{v} < \infty$$

This means that there exists two linearly independent solutions $z_1(x)$ and $z_2(x)$ of (43.14) such that

$$\int_0^\infty |z_j(x)|^2 dx < \infty \qquad (j = 1, 2)$$

On the other hand, every solution $z(x)$ of (43.14) is of the form $\gamma z_1(x) + \delta z_2(x)$. Therefore, the Minkowski inequality (20.6) yields

$$\left\{ \int_0^\infty |z(x)|^2 dx \right\}^{1/2}$$

$$= \int_0^\infty |\gamma z_1(x) + \delta z_2(x)|^2 dx \Big\}^{1/2}$$

$$\leqq |\gamma| \left\{ \int_0^\infty |z_1(x)|^2 dx \right\}^{1/2} + |\delta| \left\{ \int_0^\infty |z_2(x)|^2 dx \right\}^{1/2} < \infty$$

Next, we shall prove that, for any λ, every solution of (43.12) satisfies (43.13). To prove this, let $z_1(x)$, $z_2(x)$ be a fundamental system of the solutions of (43.14) such that

$$W_0(z_1, z_2) = 1$$

and set

$$K(x, s) = z_1(x) z_2(s) - z_1(s) z_2(x)$$

Then, for any continuous function $v(x)$,

$$u(x) = (Kv)(x) = \int_0^x K(x, s)v(s)ds$$

is a solution of the initial value problem

$$u'' + \{\lambda_0 - q(x)\}u = -v , \qquad u(0) = u'(0) = 0$$

This can be derived immediately from the fact that

$$u'(x) = \int_0^x \{z_1'(x)z_2(s) - z_1(s)z_2'(x)\}v(s)ds$$

$$u''(x) = -v(x) + \int_0^x \{(-\lambda_0+q(x))z_1(x)z_2(s)-z_1(s)(-\lambda_0+q(x))z_2(x)\}v(s)ds$$

$$= -v(x) + (-\lambda_0 + q(x))u(x)$$

Now, let $u(x, \lambda)$ be a solution of the initial value problem

(43.15) $\qquad L_xu = \lambda u , \quad u(0) = \gamma , \quad u'(0) = \delta \qquad (0 \leq x < \infty)$

Then, as was already shown in Part 5, $u(x, \lambda)$ can be expanded in a convergent series in powers of $(\lambda - \lambda_0)$

$$u(x, \lambda) = u_0(x) + (\lambda - \lambda_0)u_1(x) + (\lambda - \lambda_0)^2u_2(x) + \cdots$$

The equation (43.15) yields

$$L_xu = L_xu_0 + L_x\{(\lambda - \lambda_0)u_1 + (\lambda - \lambda_0)^2u_2 + \cdots\}$$
$$= \lambda u = \lambda u_0 + \lambda\{(\lambda - \lambda_0)u_1 + (\lambda - \lambda_0)^2u_2 + \cdots\}$$

Hence, setting $\lambda = \lambda_0$, we have

$$L_xu_0 = \lambda_0u_0 , \quad u_0(0) = \gamma , \quad u_0'(0) = \delta$$

and further

$$0 = (\lambda_0 - \lambda)u_0 + (\lambda - \lambda_0)L_xu_1 - \lambda(\lambda - \lambda_0)u_1$$
$$+ L_x\{(\lambda - \lambda_0)^2u_2 + (\lambda - \lambda_0)^3u_3 + \cdots\}$$
$$- \lambda\{(\lambda - \lambda_0)^2u_2 + (\lambda - \lambda_0)^3u_3 + \cdots\}$$

In this way, we obtain

$$L_xu_n - \lambda_0u_n = u_{n-1}, \ u_n(0) = 0, \ u_n'(0) = 0 \qquad (n \geq 1)$$

Therefore, as was mentioned above, we have

(43.16) $\qquad u_n(x) = (Ku_{n-1})(x) \qquad (n \geq 1)$

Since

$$\int_0^\infty |z_1(x)|^2dx < \infty , \qquad \int_0^\infty |z_2(x)|^2dx < \infty$$

we obtain further, by Shwarz' inequality,

$$(43.17) \qquad \int_0^\infty k(x)dx < \infty$$

where

$$k(x) = \int_0^x |K(x, s)|^2 ds$$

We shall now prove that the following inequality

$$(43.18) \qquad |u_n(x)|^2 \leq \frac{1}{n!} \int_0^\infty |u_0(x)|^2 dx \cdot \frac{d}{dx} \left\{ \int_0^x k(s)ds \right\}^n$$

holds for every $n = 1, 2, \cdots$. To prove this, we use the process of induction. Suppose that (43.18) holds for some n.

Then, by virtue of (43.16) and by making use of Schwarz' inequality,

$$|u_{n+1}(x)|^2 = \left| \int_0^x K(x, s)u_n(s)ds \right|^2 \leq \int_0^x |K(x, s)|^2 ds \int_0^x |u_n(s)|^2 ds$$

$$= k(x) \int_0^x |u_n(s)|^2 ds$$

$$\leq \frac{1}{(n + 1)!} \int_0^\infty |u_0(x)|^2 dx \cdot \frac{d}{dx} \left\{ \int_0^x k(s)ds \right\}^{n+1}$$

For the case $n = 1$, (43.18) obviously holds true and thus (43.18) holds for $n = 1, 2, \cdots$. We write

$$\int_0^\infty |u_0(x)|^2 dx = ||u_0||^2, \qquad \int_0^\infty k(x)dx = \varepsilon$$

Then from (43.18) it follows that

$$|u_n(x)|^2 \leq ||u_0||^2 \frac{\varepsilon^{n-1}}{(n-1)!} k(x)$$

This estimate, together with the Minkowski inequality (20.6), implies

$$||u(x, \lambda)|| = \left(\int_0^\infty |u(x, \lambda)|^2 dx \right)^{1/2}$$

$$\leq ||u_0|| + |\lambda - \lambda_0| ||u_1|| + |\lambda - \lambda_0| ||u_2|| + \cdots$$

$$\leq ||u_0|| \left\{ 1 + \sum_{n=1}^\infty |\lambda - \lambda_0|^n \left(\frac{\varepsilon^n}{(n-1)!} \right)^{1/2} \right\}$$

The power series on the right side converges for every λ; for its radius of convergence is

$$\frac{1}{\overline{\lim}_{n \to \infty} \left[\dfrac{\varepsilon^n}{(n-1)!} \right]^{1/2n}} = \infty$$

Thus we obtain

$$\int_0^\infty |u(x, \lambda)|^2 dx < \infty$$

q.e.d.

We shall prove the following theorem which will be used in the later paragraphs.

Theorem 43.5. All the poles and zeros of the meromorphic function $l_b(\lambda)$ are of order 1 and located on the real axis of the λ-plane.

Proof. The latter half of the theorem is already proved in Part 42. To prove the first half, suppose that $l_b(\lambda)$ has a zero of order $\geqq 2$ at a real point u_0. Then we have

$$\lim_{v \downarrow 0} \frac{\mathscr{I}(l_b(u_0 + iv))}{v} = 0$$

On the other hand, from Theorem 43.2, it follows that

$$\int_0^b |y_1(x, u_0 + iv) + l_b(u_0 + iv)y_2(x, u_0 + iv)|^2 dx = \frac{1}{v}\mathscr{I}(l_b(u_0 + iv))$$

Therefore we must have

$$\int_0^b |y_1(x, u_0) + l_b(u_0)y_2(x, u_0)|^2 dx = 0$$

This is a contradiction. Thus every zero of $l_b(\lambda)$ is of order 1. Replacing y_1 and y_2 by y_2 and y_1 respectively in the argument mentioned above, we see that every pole of $l_b(\lambda)$ is of order 1.

44. *Definition of $m_1(\lambda)$ and $m_2(\lambda)$*

So far we were concerned with the case $0 < b < \infty$. We shall also consider the case $-\infty < a < 0$. For an arbitrary real number α, the boundary condition

$$\{y_1(a, \lambda) + l_a(\lambda)y_2(a, \lambda)\} \cos \alpha + \{y_1'(a, \lambda) + l_a(\lambda)y_2'(a, \lambda)\} \sin \alpha = 0$$

at the point a determines

$$(44.1) \qquad l_a(\lambda) = -\frac{y_1(a, \lambda) \cos \alpha + y_1'(a, \lambda) \sin \alpha}{y_2(a, \lambda) \cos \alpha + y_2'(a, \lambda) \sin \alpha}$$

and also the circle $C_a(\lambda)$. Similarly as in Theorem 43.2, we can prove that, if $v = \mathscr{F}(\lambda) \neq 0$, then the fact that l lies on the circle $C_a(\lambda)$ or in its interior depends on whether

(44.2)
$$\int_a^0 |y_1(x, \lambda) + ly_2(x, \lambda)|^2 dx = -\frac{\mathscr{F}(l(\lambda))}{v}$$

or

$$\int_a^0 |y_1(x, \lambda) + ly_2(x, \lambda)|^2 dx < -\frac{\mathscr{F}(l(\lambda))}{v}$$

From this it follows that

(44.3)
$$\bigcap_{a<0} \overline{C_a(\lambda)} = C_{-\infty}(\lambda) , \qquad v = \mathscr{F}(\lambda) \neq 0$$

The point $x = -\infty$ is said to be in the limit point case or in the limit circle case, according as $C_{-\infty}(\lambda)$ is a point or a circle with non-zero finite radius. Then $x = -\infty$ is in the limit point case if and only if, for any λ, there exists at least one solution $y(x)$ of the equation

$$y'' + \{\lambda - q(x)\}y = 0 , \qquad -\infty < x \leqq 0$$

such that

$$\int_{-\infty}^0 |y(x)|^2 dx = \infty$$

Moreover, in this case, the inequality

$$\int_{-\infty}^0 |y_1(x, \lambda) + ly_2(x, \lambda)|^2 dx \leqq \frac{-\mathscr{F}(l)}{v} < \infty$$

holds for $l = C_{-\infty}(\lambda)$, $v = \mathscr{F}(\lambda) \neq 0$.

It should be noted first that $l_b(\lambda)$ is a meromorphic function of λ with simple poles on the real axis of the λ-plane and satisfies

(44.4)
$$l_b(\bar{\lambda}) = \overline{l_b(\lambda)}$$

as well as $y_1(x, \lambda)$, $y_2(x, \lambda)$, $y_1'(x, \lambda)$ and $y_2'(x, \lambda)$. In the case when $x = \infty$ is in the limit point case, we have, for $v = \mathscr{F}(\lambda) \neq 0$

(44.5)
$$\lim_{b \to \infty} l_b(\lambda) = C_\infty(\lambda) , \qquad |C_\infty(\lambda)| < \infty$$

On the other hand, the centre and the radius of $C_b(\lambda)$ are continuous function of λ for $v = \mathscr{F}(\lambda) \neq 0$, because of (43.5), (43.6) and (43.7). Therefore, Theorem 43.3 implies that, as $b \to \infty$, the $l_b(\lambda)$ is uniformly bounded on any bounded closed domain D of the λ-plane which does not meet the real axis. Hence, according to the theory of the normal family,[1] the convergence in (44.5) is uniform, so that the limit function

$$(44.6) \qquad m_2(\lambda) = \lim_{b \to \infty} l_b(\lambda)$$

is regular for $\mathscr{I}(\lambda) = v \neq 0$. In the case when $x = \infty$ is in the limit circle case, we first consider a covering of the upper and lower half planes by denumerably many bounded closed domains which do not meet the real axis. Then we can prove, similarly as above, that the theory of the normal family can be applied to $\{l_b(\lambda)\}$ in each domain. Hence we can prove, by the diagonal method, that there exists a subsequence $\{l_{b_n}(\lambda)\}$ such that

$$(44.7) \qquad b_1 < b_2 < b_3 < \cdots, \lim_{n \to \infty} b_n = \infty$$
$$\lim_{n \to \infty} l_{b_n}(\lambda) = m_2(\lambda)$$

where the convergence is uniform on any bounded closed domain in the λ-plane which does not meet the real axis. Therefore $m_2(\lambda)$ is regular for $v = \mathscr{I}(\lambda) \neq 0$.

For $x = -\infty$, we can prove similarly the following: In the limit point case,

$$(44.8) \qquad \lim_{a \to -\infty} l_a(\lambda) = m_1(\lambda)$$

and in the limit circle case, there exists a subsequence $\{l_{a_n}(\lambda)\}$ such that

$$(44.9) \qquad a_1 > a_2 > a_3 > \cdots, \lim_{n \to \infty} a_n = -\infty$$
$$\lim_{n \to \infty} l_{a_n}(\lambda) = m_1(\lambda)$$

The set $\{l_a(\lambda)\}$ in (44.8) and the sequence $\{l_{a_n}(\lambda)\}$ in (44.9) converge uniformly on any bounded closed domain of the λ-plane which does not meet the real axis, and hence $m_1(\lambda)$, in both cases, is regular for $v = \mathscr{I}(\lambda) \neq 0$.

Similarly as (44.4), we can prove

$$(44.10) \qquad l_a(\bar{\lambda}) = \overline{l_a(\lambda)}$$

From (44.4) and (44.10), it follows that

$$(44.11) \qquad m_1(\bar{\lambda}) = \overline{m_1(\lambda)}, \ m_2(\bar{\lambda}) = \overline{m_2(\lambda)}$$

Further, we obtain, by (43.10), (44.2) and the above results, that, for $v = \mathscr{I}(\lambda) \neq 0$,

[1] See Appendix, A theorem on the normal family of regular functions.

$$(44.12) \quad \int_{-\infty}^{0} |y_1(x, \lambda) + m_1(\lambda)y_2(x, \lambda)|^2 dx \leq - \mathscr{I}(m_1(\lambda))/v$$

$$\int_{0}^{\infty} |y_1(x, \lambda) + m_2(\lambda)y_2(x, \lambda)|^2 dx \leq \mathscr{I}(m_2(\lambda))/v$$

§ 2.   The General Expansion Theorem

45.   *Application of the Hilbert-Schmidt expansion theorem*

Let $[a, b]$ be an arbitrary finite closed interval. Let $f(x)$ be a *real-valued* function satisfying the following conditions:

$$(45.1)$$

$f''(x)$ is continuous on $(-\infty, \infty)$.

For some $a', b', a < a' < b' < b$,

$f(x) \equiv 0$ on the semi-infinite intervals

$-\infty < x \leq a', b' \leq x < \infty$.

Let $\{\lambda_{n,a,b}\}$ be the set of all eigenvalues of the boundary value problem

$$(45.2) \quad \begin{aligned} L_x\varphi &= \lambda\varphi \\ \varphi(a)\cos\alpha + \varphi'(a)\sin\alpha &= 0 \\ \varphi(b)\cos\beta + \varphi'(b)\sin\beta &= 0 \end{aligned}$$

and $\{\varphi_{n,a,b}\}$ be the corresponding orthonormal system of the eigenfunctions. Then, according to Theorem 21.4, $f(x)$ can be expanded in the Fourier series

$$(45.3) \quad \begin{aligned} f(x) &= \sum_{n=1}^{\infty} f_{n,a,b}\varphi_{n,a,b}(x) \\ f_{n,a,b} &= (f, \varphi_{n,a,b}) = \int_{a}^{b} f(x)\overline{\varphi_{n,a,b}(x)}dx \end{aligned}$$

which converges absolutely and uniformly on the interval $[a, b]$. In fact, since $f(x)$ satisfies the boundary conditions in (45.2) and $L_x f$ is continuous on $[a, b]$, $f(x)$ can be written as

$$(45.4) \quad f(x) = \int_{a}^{b} K(x, s)L_s f ds$$

where $K(x, s)$ is the Green's function for the boundary value problem (45.2); hence Theorem 21.4 is applicable.

Consider the inhomogeneous boundary value problem

$$L_x y - \lambda y = f(x)$$

(45.5)
$$y(a) \cos \alpha + y'(a) \sin \alpha = 0$$

$$y(b) \cos \beta + y'(b) \sin \beta = 0$$

The eigenvalues of the boundary value problem (45.2) are all real. Thus, if $\mathscr{F}(\lambda) \neq 0$, then, as was shown in Part 23, there exists a unique solution $y(x, \lambda)$ of (45.5). Moreover, $y(x, \lambda)$ can be expanded in the Fourier series

$$y(x, \lambda) = \sum_{n=1}^{\infty} C_n(\lambda) \varphi_{n,a,b}(x)$$

$$C_n(\lambda) = (y, \varphi_{n,a,b})$$

which converges absolutely and uniformly on $[a, b]$. The Fourier coefficients $C_n(\lambda)$ will be determined as follows. Since $y(x, \lambda)$ and $\varphi_{n,a,b}(x)$ satisfy the boundary conditions in (45.2), we obtain, by partial integration,

$$(L_x y, \varphi_{n,a,b}) = (y, L_x \varphi_{n,a,b})$$

Using this, we have

$$\begin{aligned}
f_{n,a,b} = (f, \varphi_{n,a,b}) &= (L_x y, \varphi_{n,a,b}) - \lambda(y, \varphi_{n,a,b}) \\
&= (y, L_x \varphi_{n,a,b}) - \lambda(y, \varphi_{n,a,b}) \\
&= (y, \lambda_{n,a,b} \varphi_{n,a,b}) - \lambda(y, \varphi_{n,a,b}) \\
&= (\lambda_{n,a,b} - \lambda) C_n(\lambda)
\end{aligned}$$

and hence

$$C_n(\lambda) = - \frac{f_{n,a,b}}{(\lambda - \lambda_{n,a,b})}$$

We thus obtain

(45.6)
$$y(x, \lambda) = \sum_{n=1}^{\infty} \frac{-f_{n,a,b} \varphi_{n,a,b}(x)}{(\lambda - \lambda_{n,a,b})}$$

On the other hand, Green's function for the boundary value problem (45.2) is of the form

(45.7)
$$\begin{aligned}
G_{a,b}(x, s, \lambda) &= -y_b(x, \lambda) y_a(s, \lambda) / W_x(y_a, y_b), \quad \text{for } x \geqq s \\
&= -y_a(x, \lambda) y_b(s, \lambda) / W_x(y_a, y_b), \quad \text{for } x < s
\end{aligned}$$

where

(45.8)
$$\begin{aligned}
y_a(x, \lambda) &= y_1(x, \lambda) + l_a(\lambda) y_2(x, \lambda) \\
y_b(x, \lambda) &= y_1(x, \lambda) + l_b(\lambda) y_2(x, \lambda)
\end{aligned}$$

Hence, by Theorem 17.1, $y(x, \lambda)$ can be written as

$$(45.9) \qquad y(x, \lambda) = \int_a^b G_{a,b}(x, s, \lambda) f(s) ds$$

Combining (45.9) with (45.6), we obtain that, if $\mathscr{F}(\lambda) \neq 0$,

$$(45.10) \qquad \sum_{n=1}^{\infty} \frac{f_{n,a,b} \varphi_{n,a,b}(x)}{\lambda - \lambda_{n,a,b}} = - \int_b^b G_{a,b}(x, s, \lambda) f(s) ds$$

From (45.3), it follows immediately that the sum of all the residues of the series in λ on the left side of (45.10) is equal to $f(x)$. This proves the following theorem.

THEOREM 45.1. The sum of all residues of the meromorphic function $- \int_a^b G_{a,b}(x, s, \lambda) f(s) ds$ in λ is equal to $f(x)$ for $a \leqq x \leqq b$.

Next, we shall calculate the residues of the function $- \int_a^b G_{a,b}(x, s, \lambda) f(s) ds$. Before the calculation, it should be noted that

$$\begin{aligned} W_x(y_a, y_b) &= W_0(y_a, y_b) = W_0(y_1 + l_a y_2, y_1 + l_b y_2) \\ &= W_0(y_1, y_1) + l_a W_0(y_2, y_1) + l_b W_0(y_1, y_2) + l_a l_b W_0(y_2, y_2) \\ &= 0 - l_a + l_b + 0 = l_b - l_a \end{aligned}$$

Hence, by virtue of (45.7), the residues under consideration are equal to those of the function

$$(45.11) \qquad \begin{aligned} & \frac{-y_b(x, \lambda)}{l_a(\lambda) - l_b(\lambda)} \int_a^x y_a(s, \lambda) f(s) ds \\ &+ \frac{-y_a(x, \lambda)}{l_a(\lambda) - l_b(\lambda)} \int_x^b y_b(s, \lambda) f(s) ds \end{aligned}$$

According to Theorem 43.5, all poles and zeros of the meromorphic functions $l_a(\lambda)$ and $l_b(\lambda)$ are of order 1 and located on the real axis of the λ-plane. Further the poles of (45.6), and hence, of (45.11), are all real and of order 1. Therefore, the pole of the function (45.11) is one of the following;

(i) λ_n such that $l_a(\lambda_n) = l_b(\lambda_n) = \mu_n \neq 0$, and $l_a(\lambda) - l_b(\lambda) \sim (\lambda - \lambda_n) \nu_n$ as $\lambda \to \lambda_n$,

(ii) λ_m' such that $l_a(\lambda_m') = l_b(\lambda_m') = 0$, and $l_a(\lambda) \sim (\lambda - \lambda_m') \mu_{1m}'$, $l_b(\lambda) \sim (\lambda - \lambda_m) \mu_{2m}'$ as $\lambda \to \lambda_m'$,

(iii) λ_k'' such that $l_a(\lambda_k'') = l_b(\lambda_k'') = \infty$, and $l_a(\lambda) \sim \mu_{1k}''(\lambda - \lambda_k'')^{-1}$, $l_b(\lambda) \sim \mu_{2k}''(\lambda - \lambda_k'')^{-1}$ as $\lambda \to \lambda_k'$.

The corresponding residues are

(i) $\dfrac{-1}{\nu_n}(y_1(x, \lambda_n) + \mu_n y_2(x, \lambda_n))((y_1(s, \lambda_n) + \mu_n y_2(s, \lambda_n)), f(s))$

(ii) $\dfrac{-1}{u'_{1m} - \mu'_{2m}}y_1(x, \lambda'_m)(y_1(s, \lambda'_m), f(s)),$

(iii) $\dfrac{-\mu''_{2k}\mu''_{1k}}{\mu''_{1k} - \mu''_{2k}}y_2(x, \lambda''_k)(y_2(s, \lambda''_k), f(s))$

respectively. Hence, by virtue of Theorem 45.1,

$$
\begin{aligned}
(45.12) \qquad f(x) = & \left\{ \sum_n - y_1(x, \lambda_n)\frac{1}{\nu_n}(y_1(s, \lambda_n), f(s)) \right.\\
& \left. + \sum_m - y_1(x, \lambda'_m)\frac{1}{\mu'_{1m} - \mu'_{2m}}(y_1(s, \lambda'_m), f(s)) \right\}\\
& + \left\{ \sum_n - y_1(x, \lambda_n)\frac{\mu_n}{\nu_n}(y_2(s, \lambda_n), f(s)) \right.\\
& \left. + \sum_n - y_2(x, \lambda_n)\frac{\mu_n}{\nu_n}(y_1(s, \lambda_n), f(s)) \right\}\\
& + \left\{ \sum_n - y_2(x, \lambda_n)\frac{\mu_n^2}{\nu_n}(y_2(s, \lambda_n), f(s)) \right.\\
& \left. + \sum_k - y_2(x, \lambda''_k)\frac{\mu''_{1k}\mu''_{2k}}{\mu''_{1k} - \mu''_{2k}}(y_2(s, \lambda''_k), f(s)) \right\}
\end{aligned}
$$

for $a \leqq x \leqq b$. On the other hand, the residues of

$$
\frac{1}{l_a(\lambda) - l_b(\lambda)}; \quad \frac{l_b(\lambda)}{l_a(\lambda) - l_b(\lambda)} \quad \text{and} \quad \frac{l_a(\lambda)}{l_a(\lambda) - l_b(\lambda)}; \quad \frac{l_a(\lambda)l_b(\lambda)}{l_a(\lambda) - l_b(\lambda)}
$$

are

$$
\frac{1}{\nu_n} \text{ (at } \lambda_n) \text{ and } \frac{1}{\mu'_{1m} - \mu'_{2m}} \text{ (at } \lambda'_m);
$$

$$
\frac{\mu_n}{\nu_n} \text{ (at } \lambda_n);
$$

$$
\frac{\mu_n^2}{\nu_n} \text{ (at } \lambda_n) \text{ and } \frac{\mu''_{1k}\mu''_{2k}}{\mu''_{1k} - \mu''_{2k}} \text{ (at } \lambda''_k)
$$

respectively. Therefore (45.12) can be written as

$$
(45.13) \quad f(x) = \int_{-\infty}^{\infty} d_u \left\{ \sum_{j,k=1}^{2} \int_0^u y_j(x, u)\left[\int_a^b f(s)y_k(s, u)ds \right]d\rho_{jk}^{(a,b)}(u) \right\}
$$

in terms of a *Stieltjes integral*, where

$$\rho_{11}^{(a,b)}(u_2) - \rho_{11}^{(a,b)}(u_1) = \frac{1}{2\pi i}\int_{C(u_1,u_2)}\frac{-d\lambda}{l_a(\lambda)-l_b(\lambda)}$$

(45.14)
$$\rho_{12}^{(a,b)}(u_2) - \rho_{12}^{(a,b)}(u_1) = \rho_{21}^{(a,b)}(u_2) - \rho_{21}^{(a,b)}(u_1)$$

$$= \frac{1}{2\pi i}\int_{C(u_1,u_2)}\frac{-l_b(\lambda)d\lambda}{l_a(\lambda)-l_b(\lambda)} = \frac{1}{2\pi i}\int_{C(u_1,u_2)}\frac{-l_a(\lambda)d\lambda}{l_a(\lambda)-l_b(\lambda)}$$

$$\rho_{22}^{(a,b)}(u_2) - \rho_{22}^{(a,b)}(u_1) = \frac{1}{2\pi i}\int_{C(u_1,u_2)}\frac{-l_a(\lambda)l_b(\lambda)d\lambda}{l_a(\lambda)-l_b(\lambda)}$$

The integrals on the right side of (45.14) are taken along the polygonal lines $C(u_1, u_2)$ connecting the points

$$u_1 - iv, u_2 - iv, u_2 + iv, u_1 + iv, u_1 - iv$$

in this order, where v is an arbitrary positive number and u_1, u_2 are arbitrary real numbers different from any one of the λ_n, λ'_m, and λ''_k.

Furthermore, on account of

$$l_a(\bar{\lambda}) = \overline{l_a(\lambda)}, \qquad l_b(\bar{\lambda}) = \overline{l_b(\lambda)}$$

and the arbitrariness of v in (45.14), (45.14) may be written as

$$\rho_{jk}^{(a,b)}(u_2) - \rho_{jk}^{(a,b)}(u_1) = \lim_{v\downarrow 0}\frac{1}{\pi}\int_{u_1}^{u_2}f_{jk}^{(a,b)}(u+iv)du$$

(45.15)
$$f_{11}^{(a,b)}(\lambda) = \mathscr{I}\frac{1}{l_a(\lambda)-l_b(\lambda)}, \quad f_{12}^{(a,b)}(\lambda) = f_{21}^{(a,b)}(\lambda)$$

$$= \mathscr{I}\frac{l_b(\lambda)}{l_a(\lambda)-l_b(\lambda)} \left(\text{as well as } \mathscr{I}\frac{l_a(\lambda)}{l_a(\lambda)-l_b(\lambda)}\right)$$

$$f_{12}^{(a,b)}(\lambda) = \mathscr{I}\frac{l_a(\lambda)l_b(\lambda)}{l_a(\lambda)-l_b(\lambda)}$$

46. Helly's theorem and Poisson's integral formula

We shall start with the *Helly selection theorem*, which will be needed in the following.

THEOREM 46.1. Let $\{v_n(x)\}$, $n=1, 2, \cdots$, be a sequence of monotone increasing functions on $(-\infty, \infty)$. If

(46.1) $$\sup_{n\geq 1, \ -\infty < x < \infty}|v_n(x)| < \infty$$

then there exists a subsequences $\{v_{n'}(x)\}$ such that

$$\lim_{n'\to\infty}v_{n'}(x) = v_\infty(x)$$

exists for every x. Further the limit function $v_\infty(x)$ is monotone increasing.

We shall first prove the following

Lemma. Let $v(x)$ be monotone increasing and bounded. Then the set of all discontinuity points of $v(x)$ consists of at most denumerably infinite points.

Proof. Let $t_1, t_2, \cdots, t_k$ be arbitrary numbers such that

$$-\infty < t_1 < t_2 < \cdots < t_k < \infty$$

Then

$$v(\infty) - v(-\infty) = \sum_{j=0}^{k} \{v(t_{j+1}) - v(t_j)\}$$

where $t_0 = -\infty$, $t_{k+1} = \infty$. We denote by

$$v(t' + 0) - v(t' - 0)$$

the jump of $v(x)$ at a discontinuity point t' and further by A_n the set of all discontinuity points with jump $\geq 1/n$. Obviously, any discontinuity point is contained in some A_n. Moreover, for any finite number of discontinuity points $t_1', t_2', \cdots, t_k'$,

$$v(\infty) - v(-\infty) \geq \sum_{j=1}^{k} \{v(t_j' + 0) - v(t_j' - 0)\}$$

From this it follows that the number of all points in A_n is less than $n\{v(\infty) - v(-\infty)\}$, and hence, finite. Thus the proof is completed.

Proof of Theorem 46.1. Since the set of all rational numbers is denumerable, we can arrange it as $x_1, x_2, \cdots$. On account of (46.1), the sequence of numbers $\{v_n(x_1)\}$ is bounded. Hence there exists a convergent subsequence

$$v_{1(1)}(x_1), v_{2(1)}(x_1), v_{3(1)}(x_1), \cdots$$

Further, on account of (46.1), the sequence of numbers $\{v_{n(1)}(x_2)\}$ is also bounded. Hence there exists a convergent subsequence

$$v_{1(2)}(x_2), v_{2(2)}(x_2), v_{3(2)}(x_2), \cdots$$

Repeating this procedure, we finally select a subsequence

(46.2) $v_{1(1)}(x), v_{2(2)}(x), v_{3(3)}(x), \cdots, v_{n(n)}(x), \cdots$

from the original sequence of functions $\{v_n(x)\}$ (diagonal method). Then (46.2) converges for every rational number $x = x_1, x_2, x_3, \cdots$. We denote (46.2) by $\{v_{n''}(x)\}$, and set

$$C_j = \lim_{n'' \to \infty} v_{n''}(x_j) \qquad (j = 1, 2, \cdots)$$

Obviously, if $x_k < x_m$,

$$C_k \leqq C_m$$

Accordingly, if we set

$$(46.3) \qquad v(x) = \inf_{x_j > x} C_j$$

then $v(x)$ is monotone increasing and, by (46.1), bounded. Further, from

$$\inf_{x_j > x} C_j = \lim_{\varepsilon \downarrow 0} \inf_{x_k > x+\varepsilon} C_k$$

it follows that

$$(46.4) \qquad v(x) = v(x + 0) = \lim_{\varepsilon \downarrow 0} v(x + \varepsilon)$$

The set of all rational numbers is dense in the set of all real numbers. Hence, for any $\varepsilon > 0$ and for any real number x, there exist rational numbers x_j, x_k such that $x > x_j > x - \varepsilon > x_k > x - 2\varepsilon$. Hence we obtain

$$v(x) \geqq C_j \geqq v(x - \varepsilon) \geqq C_k \geqq v(x - 2\varepsilon)$$

and thus

$$(46.5) \qquad v(x) \geqq v(x - 0) = \lim_{\varepsilon \downarrow 0} v(x - \varepsilon) = \sup_{x_j < x} C_j$$

If $x < x_j$, then $v_n(x) \leqq v_n(x_j)$, and hence $\overline{\lim}_{n'' \to \infty} v_{n''}(x) \leqq C_j$. This, together with (46.3) and (46.4), implies

$$(46.6) \qquad \overline{\lim}_{n'' \to \infty} v_{n''}(x) \leqq \inf_{x_j > x} C_j = v(x) = v(x + 0)$$

Similarly, if $x > x_j$, then $C_j \leqq \underline{\lim}_{n'' \to \infty} v_{n''}(x)$, and hence, by (46.5),

$$(46.7) \qquad \sup_{x_j < x} C_j = v(x - 0) \leqq \underline{\lim}_{n'' \to \infty} v_{n''}(x)$$

From (46.6) and (46.7), it follows that, at every continuity point x of $v(x)$, for which $v(x - 0) = v(x + 0)$ holds, $\lim_{n'' \to \infty} v_{n''}(x)$ exists and equals $v(x)$.

According to the lemma, the set of all discontinuity points of the bounded monotone increasing function $v(x)$ consists of at most denumerably infinite points. Hence the set of all points x for which $\lim_{n'' \to \infty} v_{n''}(x)$ does not exist consists of at most denumerably infinite points and may be arranged as $x_1', x_2', \cdots$. Applying the diagonal method to $\{x_k'\}$ and $\{v_{n''}(x)\}$, we can find a subsequence $\{v_{n'}(x)\}$ of $\{v_{n''}(x)\}$ such that

$$\lim_{n' \to \infty} v_{n'}(x_j') \qquad\qquad (j = 1, 2, \cdots)$$

exists. Thus the limit function $v_\infty(x)$ of $\{v_{n'}(x)\}$ exists and is a bounded monotone increasing function, q.e.d.

According to the Jordan decomposition theorem, if $v(x)$ is of bounded variation, then

$$v(x) = p(x) - n(x)$$

where $p(x)$ and $n(x)$ are monotone increasing functions. Further, both $p(\infty) - p(-\infty)$ and $n(\infty) - n(-\infty)$ do not exceed the total variation $\int_{-\infty}^{\infty} d|v|(x)$ of $v(x)$. Accordingly, Theorem 46.1 implies the following

COROLLARY. Let $\{v_n(x)\}$ be a sequence of functions of bounded variation. If, moreover,

$$(46.8) \qquad v_n(-\infty) = \text{constant} \qquad (n = 1, 2, \cdots)$$

$$(46.9) \qquad \sup_{n \geq 1} \left\{ \int_{-\infty}^{\infty} d|v_n|(x) \right\} < \infty$$

then there exists a subsequence $\{v_{n'}(x)\}$ such that, for every x,

$$(46.10) \qquad \lim_{n' \to \infty} v_{n'}(x) = v_\infty(x)$$

exists and the limit function $v_\infty(x)$ is of bounded variation.

Together with Theorem 46.1, the following theorem plays an important role in the following.

THEOREM 46.2. Let $\{v_{n'}(x)\}$ be a sequence of functions as in the corollary, and $v_\infty(x) = \lim_{n' \to \infty} v_{n'}(x)$. Let us assume further that, for any $\varepsilon > 0$ and $M > 0$, there exists $N > 0$ such that

$$(46.9') \qquad \int_{|x| \geq N} d|v_{n'}|(x) < \varepsilon M^{-1}, \qquad \int_{|x| \geq N} d|v_\infty|(x) < \varepsilon M^{-1}$$

for every n'. Then, for every bounded continuous function $f(x)$ on $(-\infty, \infty)$,

$$(46.11) \qquad \lim_{n' \to \infty} \int_{-\infty}^{\infty} f(x) dv_{n'}(x) = \int_{-\infty}^{\infty} f(x) dv_\infty(x)$$

Proof. Put $\sup_{-\infty < x < \infty} |f(x)| = M$. Then, by virtue of $(46.9')$,

$$\left| \int_{-\infty}^{\infty} f(x) dv_{n'}(x) - \int_{-N}^{N} f(x) dv_{n'}(x) \right| \leq M \frac{\varepsilon}{M} = \varepsilon$$

$$\left| \int_{-\infty}^{\infty} f(x) dv_\infty(x) - \int_{-N}^{N} f(x) dv_\infty(x) \right| \leq M \frac{\varepsilon}{M} = \varepsilon$$

Hence it is sufficient to prove that

$$(46.12) \qquad \lim_{n' \to \infty} \int_{-N}^{N} f(x) dv_{n'}(x) = \int_{-N}^{N} f(x) dv_\infty(x)$$

To prove this, divide the interval $[-N, N]$ into $k - 1$ equal parts,

$$-N = x_1 < x_2 < \cdots < x_k = N$$

and put

$$f_k(x) = f(x_j) \quad \text{for} \quad x_j \leq x < x_{j+1} \qquad (j = 1, 2, \cdots, k - 1)$$

Since $f(x)$ is continuous on the finite closed interval $[-N, N]$, it is uniformly continuous. Hence

$$(46.13) \qquad \lim_{k \to \infty} f_k(x) = f(x) \text{ uniformly on } [-N, N)$$

From (46.13) and (46.9) it follows that

$$\lim_{k \to \infty} \int_{-N}^{N} f_k(x) dv_{n'}(x) = \int_{-N}^{N} f(x) dv_{n'}(x)$$

uniformly with respect to $n' = 1', 2', \cdots, \infty$. Consequently, it is sufficient to prove that, for the *step function* $f_k(x)$,

$$(46.12') \qquad \lim_{n' \to \infty} \int_{-N}^{N} f_k(x) dv_{n'}(x) = \int_{-N}^{N} f_k(x) dv_\infty(x)$$

For the step function $f_k(x)$, we have

$$\int_{-N}^{N} f_k(x) dv_{n'}(x) = \sum_{j=1}^{k-1} f(x_j)[v_{n'}(x_{j+1}) - v_{n'}(x_j)]$$

$$\int_{-N}^{N} f_k(x) dv_\infty(x) = \sum_{j=1}^{k-1} f(x_j)[v_\infty(x_{j+1}) - v_\infty(x_j)]$$

Hence (46.10) implies (42.12'), q.e.d.

Next we shall prove the following theorem.

THEOREM 46.3. Let $h(z)$ be harmonic in the domain $|z| < 1$ of the complex z-plane, and let $h(z)$ satisfy the following condition

$$(46.14) \qquad \frac{1}{2\pi} \int_{-\pi}^{\psi} |h(re^{i\theta})| d\theta = \tau_r(\psi) \leq c < \infty$$

for $-\pi \leq \psi \leq \pi$ and $0 \leq r < 1$.

Then there exists a function $\tau(\psi)$ of bounded variation on the closed interval $[-\pi, \pi]$ such that

$$(46.15) \qquad h(z) = h(re^{i\theta}) = \int_{-\pi}^{\pi} \mathscr{R} \frac{e^{i\psi} + re^{i\theta}}{e^{i\psi} - re^{i\theta}} d\tau(\psi) \qquad \text{for } 0 \leq r < 1,$$

where the total variation of $\tau(\psi)$ satisfies

$$\int_{-\pi}^{\pi} d\,|\tau(\psi)| \leqq c$$

Proof. Put

$$h_n(z) = h\left(\frac{n-1}{n}z\right) \qquad (n = 1, 2, \cdots)$$

Then $h_n(z)$ is harmonic and continuous on the domain $|z| < n/(n-1)$ which contains the closed domain $|z| \leqq 1$ in its interior. Hence, by the *Poisson integral formula*[1] we obtain

$$h_n(re^{i\theta}) = \frac{1}{2\pi}\int_{-\pi}^{\pi}\frac{1-r^2}{1-2r\cos(\theta-\psi)+r^2}\mu_n(\psi)d\psi \qquad (0 \leqq r < 1)$$

where $\mu_n(\theta) = h_n(e^{i\theta})$. Put

(46.16) $$\tau^{(n)}(\psi) = \frac{1}{2\pi}\int_{-\pi}^{\psi}\mu_n(\varphi)d\varphi$$

Then, on account of

$$\frac{1-r^2}{1-2r\cos(\theta-\psi)+r^2} = \mathscr{R}\frac{e^{i\psi}+re^{i\theta}}{e^{i\psi}-re^{i\theta}}$$

we have

(46.17) $$h_n(re^{i\theta}) = \int_{-\pi}^{\pi}\mathscr{R}\frac{e^{i\psi}+re^{i\theta}}{e^{i\psi}-re^{i\theta}}d\tau^{(n)}(\psi) \qquad (0 \leqq r < 1)$$

On the other hand, according to (46.14),

$$\frac{1}{2\pi}\int_{-\pi}^{\psi}|\mu_n(\varphi)|\,d\varphi = \tau_{(n-1)/n}(\psi) \leqq c \qquad (n = 1, 2, \cdots)$$

Hence every $\tau^{(n)}(\psi)$ is of bounded variation on $[-\pi, \pi]$ and

(46.18) $\tau^{(n)}(-\pi) = 0$, and

the (total variation of $\tau^{(n)}(\psi)$) $= \displaystyle\int_{-\pi}^{\pi} d\,|\tau^{(n)}|(\psi) \leqq c$.

Moreover $\lim_{n\to\infty} h_n(re^{i\theta}) = h(re^{i\theta})$ for $0 \leqq r < 1$. Therefore, Theorem 46.2, together with Theorem 46.1, implies that there exists a subsequence $\{n'\}$ of $\{n\}$ such that $\lim_{n'\to\infty}\tau^{(n')}(\psi) = \tau(\psi)$ exist and for which (46.15) holds, q.e.d.

[1] See Appendix, Poisson integral formula.

47. *The Weyl-Stone-Titchmarsh-Kodaira theorem*

We shall first prove the following

THEOREM 47.1. Let $f_{jk}^{(a,b)}(\lambda)$ be as in (45.15). If $v = \mathscr{I}(\lambda) > 0$, then[1]

$$f_{11}^{(a,b)}(\lambda) = v \int_a^b |G_{a,b}(0, s, \lambda)|^2 ds$$

(47.1)
$$f_{12}^{(a,b)}(\lambda) = f_{21}^{(a,b)}(\lambda) = v \int_a^b G_{a,b}(0, s, \lambda)\overline{G'_{a,b}(0, s, \lambda)}ds$$

$$f_{22}^{(a,b)}(\lambda) = v \int_a^b |G'_{a,b}(0, s, \lambda)|^2 ds$$

Hence, in particular,

(47.2) $\qquad f_{11}^{(a,b)}(\lambda) \geqq 0, f_{22}^{(a,b)}(\lambda) \geqq 0$ for $\mathscr{I}(\lambda) = v > 0$

and further

(47.3) $\qquad |f_{12}^{(a,b)}(\lambda)|^2 \leqq f_{11}^{(a,b)}(\lambda)f_{22}^{(a,b)}(\lambda)$ for $\mathscr{I}(\lambda) = v > 0$

Proof. From (42.2) and (45.8), it follows that

$$W_x(y_a, y_b) = -l_a(\lambda) + l_b(\lambda)$$

Hence, from (42.2), (45.7), and (45.8), it follows that

(47.4)
$$G_{a,b}(0, s, \lambda) = \begin{cases} \dfrac{y_a(s, \lambda)}{l_a(\lambda) - l_b(\lambda)} & (0 \geqq s) \\[2ex] \dfrac{y_b(s, \lambda)}{l_a(\lambda) - l_b(\lambda)} & (0 < s) \end{cases}$$

and

(47.5)
$$G'_{a,b}(0, s, \lambda) = \begin{cases} \dfrac{l_b(\lambda)y_a(s, \lambda)}{l_a(\lambda) - l_b(\lambda)} & (0 \geqq s) \\[2ex] \dfrac{l_a(\lambda)y_b(s, \lambda)}{l_a(\lambda) - l_b(\lambda)} & (0 < s) \end{cases}$$

Therefore, by (43.10′), (44.2), and (45.15), we obtain

$$\int_a^b |G_{a,b}(0, s, \lambda)|^2 ds = \frac{-\mathscr{I}(l_a(\lambda)) + \mathscr{I}(l_b(\lambda))}{v|l_a(\lambda) - l_b(\lambda)|^2}$$

$$= \frac{1}{v} \mathscr{I} \frac{1}{l_a(\lambda) - l_b(\lambda)} = \frac{1}{v}f_{11}^{(a,b)}(\lambda)$$

$$\int_a^b |G'_{a,b}(0, s, \lambda)|^2 ds = \frac{-|l_b(\lambda)|^2 \mathscr{I}(l_a(\lambda)) + |l_a(\lambda)|^2 \mathscr{I}(l_b(\lambda))}{v|l_a(\lambda) - l_b(\lambda)|^2}$$

$$= \frac{1}{v} \mathscr{I} \frac{l_a(\lambda)l_b(\lambda)}{l_a(\lambda) - l_b(\lambda)} = \frac{1}{v}f_{22}^{(a,b)}(\lambda)$$

---

[1] $G'_{a,b}(0, s, \lambda)$ means $[\partial/\partial x G_{a,b}(x, s, \lambda)]_{x=0}$.

and

$$\int_a^b G_{a,b}(0, s, \lambda)\overline{G'_{a,b}(0, s, \lambda)}ds = \frac{-\overline{l_b(\lambda)}\,\mathscr{F}\,(l_a(\lambda)) + \overline{l_a(\lambda)}\,\mathscr{F}\,(l_b(\lambda))}{v\,|\,l_a(\lambda) - l_b(\lambda)\,|^2}$$

$$= \frac{1}{v}\,\mathscr{F}\,\frac{l_b(\lambda)}{l_a(\lambda) - l_b(\lambda)} = \frac{1}{v}f_{12}^{(a,b)}(\lambda) = \frac{1}{v}f_{21}^{(a,b)}(\lambda)$$

The equation (47.3) can be derived immediately from (47.1), by making use of the Schwarz inequality (20.5), q.e.d.

The linear transformation

(47.6) $$\lambda = u + iv = i(1 - z)/(1 + z)$$

maps one to one and conformally the upper half plane $v = \mathscr{F}(\lambda) > 0$ of the λ-plane onto the interior $|z| < 1$ of the unite circle of the z-plane. We write the domain $|z| < 1$ as

(47.7) $$z = re^{i\theta}, \qquad 0 \le r < 1$$

Then we shall prove that, in the domain (47.7),

(47.8) $$f_{jk}^{(a,b)}(\lambda) = f_{jk}^{(a,b)}\left(i\frac{1 - z}{1 + z}\right) = \int_{-\pi}^{\pi} \mathscr{R}\frac{e^{i\psi} + re^{i\theta}}{e^{i\psi} - re^{i\theta}}d\tau_{jk}^{(a,b)}(\psi)$$

for $j, k = 1, 2$. To prove this, it is sufficient to prove that $f_{jk}^{(a,b)}(i(1 - z)/(1 + z))$ satisfies (46.14). Since $f_{jj}^{(a,b)}(i(1 - z)/(1 + z))$ is harmonic and ≥ 0 in the domain $|z| < 1$,

$$\frac{1}{2\pi}\int_{-\pi}^{\pi}\left|f_{jj}^{(a,b)}\left(i\frac{1 - re^{i\theta}}{1 + re^{i\theta}}\right)\right|d\theta = \frac{1}{2\pi}\int_{-\pi}^{\pi}f_{jj}^{(a,b)}\left(i\frac{1 - re^{i\theta}}{1 + re^{i\theta}}\right)d\theta = f_{jj}^{(a,b)}(i)^1$$

This, together with (47.2) and (47.3), implies

$$\frac{1}{2\pi}\int_{-\pi}^{\pi}\left|f_{12}^{(a,b)}\left(i\frac{1 - re^{i\theta}}{1 + re^{i\theta}}\right)\right|d\theta = \frac{1}{2\pi}\int_{-\pi}^{\pi}\left|f_{21}^{(a,b)}\left(i\frac{1 - re^{i\theta}}{1 + re^{i\theta}}\right)\right|d\theta$$

$$\le \frac{1}{2\pi}\left\{\int_{-\pi}^{\pi}f_{11}^{(a,b)}\left(i\frac{1 - re^{i\theta}}{1 + re^{i\theta}}\right)d\theta\int_{-\pi}^{\pi}f_{22}^{(a,b)}\left(i\frac{1 - re^{i\theta}}{1 + re^{i\theta}}\right)d\theta\right\}^{1/2}$$

$$= \{f_{11}^{(a,b)}(i)f_{22}^{(a,b)}(i)\}^{1/2}$$

Accordingly, Theorem 46.3 implies (47.8).

It is easily seen that $\tau_{jk}^{(a,b)}(\psi)$ in (47.8) statisfies the following conditions:

[1] If $h(re^{i\theta})$ is harmonic for $0 \le r < 1$, then

$$h(0) = \frac{1}{2\pi}\int_{-\pi}^{\pi}h(re^{i\theta})d\theta$$

$\tau_{jj}^{(a,b)}(\psi)$ is monotone increasing on $[-\pi, \pi]$,

$\tau_{jj}^{(a,b)}(-\pi) = 0$, $\tau_{jj}^{(a,b)}(\pi) = f_{jj}^{(a,b)}(i)$,

(47.9) $\qquad \tau_{jk}^{(a,b)}(\psi) = \tau_{kj}^{(a,b)}(\psi)$, $j \neq k$, is of bounded

variation on $[-\pi, \pi]$, $\tau_{jk}^{(a,b)}(-\pi) = 0$,

$$\int_{-\pi}^{\pi} d \, |\, \tau_{jk}^{(a,b)}\, |\, (\psi) \leqq \{f_{11}^{(a,b)}(i) f_{22}^{(a,b)}(i)\}$$

Now we shall return to the z-plane. From (47.6) we obtain

(47.10) $$z = re^{i\theta} = \frac{i - \lambda}{i + \lambda}, \qquad \lambda = u + iv$$

Substituting $e^{i\psi}$, s for z, λ in (47.6), we have

(47.11) $$s = i\frac{1 - e^{i\psi}}{1 + e^{i\psi}} = \tan\frac{\psi}{2}$$

Hence the unite circle $e^{i\psi}$, $-\pi \leqq \psi \leqq \pi$, of the z-plane corresponds to the real-axis s, $-\infty < s < \infty$, of the λ-plane under the transformation (47.11). Accordingly, by (47.10),

(47.12)
$$\mathscr{R}\frac{e^{i\psi} + re^{i\theta}}{e^{i\psi} - re^{i\theta}} = \mathscr{R}\frac{e^{i\psi} + (i - \lambda)/(i + \lambda)}{e^{i\psi} - (i - \lambda)/(i + \lambda)} = \mathscr{R}\frac{i(\lambda s + 1)}{\lambda - s}$$
$$= \mathscr{R}\frac{i\,|\,\lambda\,|^2 s - i\lambda s^2 + i(\bar{\lambda} - s)}{|\,\lambda - s\,|^2} = \frac{v(s^2 + 1)}{(u - s)^2 + v^2}$$

Consequently, we obtain, by (47.8) and (47.11), that

(47.13)
$$f_{jk}^{(a,b)}(u + iv) = \int_{-\infty}^{\infty} \frac{v(s^2 + 1)}{(u - s)^2 + v^2} d\tau_{jk}^{(a,b)}(2\tan^{-1} s)$$
$$+ v\{\tau_{jk}^{(a,b)}(\pi) - \tau_{jk}^{(a,b)}(\pi - 0) + \tau_{jk}^{(a,b)}(-\pi + 0) - \tau_{jk}^{(a,b)}(-\pi)\}$$

for $v > 0$.

Next we shall prove that

(47.14) $$\lim_{a = a_n \to -\infty,\, b = b_n \to \infty} f_{jk}^{(a,b)}(\lambda) = f_{jk}(\lambda)$$

uniformly on any bounded closed domain D which does not meet the real axis of the λ-plane, and moreover that

(47.15)
$$f_{11}(\lambda) = \mathscr{S}\frac{1}{m_1(\lambda) - m_2(\lambda)}, \quad f_{12}(\lambda) = f_{21}(\lambda)$$
$$= \mathscr{S}\frac{m_2(\lambda)}{m_1(\lambda) - m_2(\lambda)} \left(\text{as well as } \mathscr{S}\frac{m_1(\lambda)}{m_1(\lambda) - m_2(\lambda)}\right)$$
$$f_{22}(\lambda) = \mathscr{S}\frac{m_1(\lambda)m_2(\lambda)}{m_1(\lambda) - m_2(\lambda)}$$

As was shown in (44.7) and (44.9), $l_{a_n}(\lambda)$ and $l_{b_n}(\lambda)$ converge uniformly to $m_1(\lambda)$ and $m_2(\lambda)$ on the domain D, respectively. Hence, in order to prove (47.14) and (47.15), it is sufficient to show that $m_1(\lambda) \neq m_2(\lambda)$ for any λ in D. Suppose first $m_1(\lambda) \equiv m_2(\lambda)$. Then from (44.12) it follows that

$$\int_{-\infty}^{\infty} |y_1(x, \lambda) + m_1(\lambda)y_2(x, \lambda)|^2 dx \leq 0$$

This is a contradiction, and hence $m_1(\lambda) \not\equiv m_2(\lambda)$. Suppose next $m_1(\lambda_0) = m_2(\lambda_0)$ for some λ_0 in D. Then, by virtue of (44.7) and (44.9), Hurwitz's theorem[1] implies that, for any $\varepsilon > 0$, there exist $l_{a_n}(\lambda)$, $l_{b_n}(\lambda)$ and λ_n such that

$$|\lambda_n - \lambda_0| < \varepsilon \text{ and } l_{a_n}(\lambda_n) = l_{b_n}(\lambda_n)$$

Since $\mathscr{I}(\lambda_0) \neq 0$, we can choose λ_n with $\mathscr{I}(\lambda_n) \neq 0$ by taking ε sufficiently small. Hence there exists a non real zero λ_n of

$$W_x(y_{a_n}, y_{b_n}) = -l_{a_n}(\lambda) + l_{b_n}(\lambda)$$

This contradicts the fact mentioned in Part 45 (p. 176), q.e.d.

Since $\tau_{jk}^{(a,b)}(\psi)$ satisfies (47.9), we obtain, by Theorem 46.1, a sub-sequence $\{n'\}$ of $\{n\}$ such that

$$(47.16) \qquad \lim_{a=a_{n'} \to -\infty, b=b_{n'} \to \infty} \tau_{jk}^{(a,b)}(2 \tan^{-1} s) = \tau_{jk}(2 \tan^{-1} s) < \infty$$

exists for every s, $s = -\infty$ and $s = \infty$ inclusive, and moreover, on account of (47.14),

$$(47.17) \qquad f_{jk}(u + iv) = \int_{-\infty}^{\infty} \frac{v(s^2 + 1)}{(u - s)^2 + v^2} d\tau_{jk}(2 \tan^{-1} s)$$
$$+ v \lim_{a=a_{n'} \to -\infty, b=b_{n'} \to \infty} \{\tau_{jk}^{(a,b)}|_{\pi-0}^{\pi} - \tau_{jk}^{(a,b)}|_{-\pi}^{-\pi+0}\}$$

for $v > 0$.

On the other hand, from (45.15) and (47.13), it follows that

$$\rho_{jk}^{(a,b)}(u_2) - \rho_{jk}^{(a,b)}(u_1)$$
$$= \lim_{v \downarrow 0} \frac{1}{\pi} \int_{u_1}^{u_2} du \left\{\int_{\infty}^{\infty} \frac{v(s^2 + 1)}{(u - s)^2 + v^2} d\tau_{jk}^{(a,b)}(2 \tan^{-1} s)\right\}$$
$$= \int_{u_1}^{u_2} (s^2 + 1) d\tau_{jk}^{(a,b)}(2 \tan^{-1} s)$$

for every pair of the continuity points $s = u_1, u_2$ of $\tau_{jk}^{(a,b)}(2 \tan^{-1} s)$. For

---

[1] See Appendix, Hurwitz's theorem

$$\lim_{v \downarrow 0} \int_{u_1}^{u_2} \frac{v\,du}{(u-s)^2 + v^2} = \lim_{v \downarrow 0} \left[\tan^{-1} \frac{u-s}{v} \right]_{u_1}^{u_2}$$

$$= \pi \text{ for } u_1 < s < u_2$$

$$= \frac{\pi}{2} \text{ for } s = u_2$$

$$= \frac{\pi}{2} \text{ for } s = u_1$$

$$= 0 \text{ for } s < u_1, \text{ and } s > u_2$$

and the non-negative function

$$(s^2 + 1)\frac{v}{(u-s)^2 + v^2}$$

is integrable, that is,

$$\int_{u_1}^{u_2} \int_{-\infty}^{\infty} \frac{v(s^2+1)}{(u-s)^2+v^2} du\, d\tau_{jk}^{(a,b)}(2\tan^{-1}s) < \infty$$

The latter property allows the interchange of the order of integration in the above calculation (Fubini's theorem), and by the continuity at $s = u_1$ and u_2 of $\tau_{jk}^{(a,b)}(2\tan^{-1} s)$, the discontinuity of the function

$$\lim_{v \downarrow 0} \frac{1}{\pi} \int_{u_1}^{u_2} \frac{v\,du}{(u-s)^2 + v^2}$$

is irrelevant in the above calculation.

Accordingly, if both $s = u_1$ and $s = u_2$ are continuity points of $\tau_{jk}^{(a,b)}(2\tan^{-1} s)$, $(a = a_{n'}, b = b_{n'}; n = 1, 2, \cdots)$, and of $\tau_{jk}(2\tan^{-1} s)$, then Theorem 46.2, together with (47.15)–(47.17), implies

(47.18)
$$\lim_{a = a_{n'} \to -\infty,\, b = b_{n'} \to \infty} \{ \rho_{jk}^{(a,b)}(u_2) - \rho_{jk}^{(a,b)}(u_1) \}$$
$$= \int_{u_1}^{u_2} (s^2 + 1)d\tau_{jk}(2\tan^{-1} s)$$
$$= \lim_{v \downarrow 0} \frac{1}{\pi} \int_{u_1}^{u_2} du \left\{ \int_{-\infty}^{\infty} \frac{v(s^2+1)}{(u-s)^2+v^2} d\tau_{jk}(2\tan^{-1} s) \right\}$$
$$= \lim_{v \downarrow 0} \int_{u_1}^{u_2} f_{jk}(u+iv)du = \rho_{jk}(u_2) - \rho_{jk}(u_1)$$

According to (47.2) and (47.3), for $v = \mathscr{I}(\lambda) > 0$,

$$f_{11}(\lambda) \geqq 0, f_{22}(\lambda) \geqq 0, |f_{12}(\lambda)|^2 \leqq f_{11}(\lambda)f_{22}(\lambda)$$

Hence by making use of the Schwarz inequality, we obtain, for $u_2 > u_1$,

$$\rho_{11}(u_2) - \rho_{11}(u_1) \geqq 0 , \qquad \rho_{22}(u_2) - \rho_{22}(u_1) \geqq 0 ,$$

$$\rho_{12}(u_2) - \rho_{12}(u_1) = \rho_{21}(u_2) - \rho_{21}(u_1) ,$$

$$(\rho_{11}(u_2) - \rho_{11}(u_1))(\rho_{22}(u_2) - \rho_{22}(u_1)) \geqq | \rho_{12}(u_2) - \rho_{12}(u_1) |^2$$

This proves that the *density matrix*

(47.19)
$$\begin{bmatrix} \rho_{11}(u_2) - \rho_{11}(u_1) , & \rho_{12}(u_2) - \rho_{12}(u_1) \\ \rho_{21}(u_2) - \rho_{21}(u_1) , & \rho_{22}(u_2) - \rho_{22}(u_1) \end{bmatrix}$$

is *symmetric* and *positive definite* for $u_2 > u_1$.

Now we shall prove the general expansion theorem as follows. To begin with, we recall that, whenever $f(x)$ is a function with the property (45.1), the equation (45.13) is valid and the integral in (45.13) converges absolutely and uniformly on the interval $[a, b]$. Multiply both sides of (45.13) by $f(x)$ and integrate from a to b. Then, by the fact mentioned above, we can change the order of integration and obtain

$$(f,f) = \int_{-\infty}^{\infty} du \left\{ \sum_{j,k=1}^{2} \int_{0}^{u} (f(x), y_j(x, u)) d\rho_{jk}^{(a,b)}(u)(f(s), y_k(s, u)) \right\}$$

We write this as follows:

(47.20)
$$(f,f) = \int_{-\infty}^{\infty} du \left\{ \sum_{j,k=1}^{2} \int_{0}^{u} f_j(u) d\rho_{jk}^{(a,b)}(u) f_k(u) \right\}$$

$$f_j(u) = (f(x), y_j(x, u))$$

For a function $g(x)$ with the property (45.1), we also obtain

(47.21)
$$(g,g) = \int_{-\infty}^{\infty} du \left\{ \sum_{j,k=1}^{2} \int_{0}^{u} g_j(u) d\rho_{jk}^{(a,b)}(u) g_k(u) \right\}$$

$$g_j(u) = (g(x), y_j(x, u))$$

Substituting $f + \nu g$, ν being an arbitrary parameter, for f in (47.20), we obtain

(47.22)
$$(g,f) = \int_{-\infty}^{\infty} du \left\{ \sum_{j,k=1}^{2} \int_{0}^{u} g_j(u) d\rho_{jk}^{(a,b)}(u) f_k(u) \right\}$$

We denote by $F(u)$, $G(u)$, $H(u)$ the functions in brackets in (47.20), (47.21) and (47.22), respectively.

We use the following facts. If a matrix $(\sigma_{jk})_{j,k=1,2}$ is symmetric and positive definite, then, for any real numbers ξ_j, ξ_k,

(47.23)
$$\sum_{j,k=1}^{2} \xi_j \sigma_{jk} \xi_k \geqq 0^1$$

---

^{1} $\sum_{j,k=1}^{2} \xi_j \sigma_{jk} \xi_k = (\sqrt{\sigma_{11}} \cdot \xi_1 - \sqrt{\sigma_{22}} \cdot \xi_2)^2 + 2\sqrt{\sigma_{11}\sigma_{22}} \cdot \xi_1 \xi_2 + 2\sigma_{12} \xi_1 \xi_2$
$\geqq (\sqrt{\sigma_{11}} \cdot \xi_1 - \sqrt{\sigma_{22}} \cdot \xi_2)^2 + 2(\sqrt{\sigma_{11}\sigma_{22}} - |\sigma_{12}|)\xi_1 \xi_2 \geqq 0$

Substituting $\xi_j + \varepsilon\eta_j$, ε being a real parameter, for ξ_j in (47.23), we obtain

$$\sum_{j,k=1}^{2} (\xi_j + \varepsilon\eta_j)\sigma_{jk}(\xi_k + \varepsilon\eta_k) \geq 0$$

Hence, by considering this as an equation of degree 2 in unknown ε, we see that the discriminant of the equation must be non-positive, that is,

$$(47.24) \qquad \left| \sum_{j,k=1}^{2} \xi_j \sigma_{jk}\eta_k \right| \leq \left\{ \left(\sum_{j,k=1}^{2} \xi_j\sigma_{jk}\xi_k \right)\left(\sum_{j,k=1}^{2} \eta_j\sigma_{jk}\eta_k \right) \right\}^{1/2}$$

Since the density matrix $[d\rho_{jk}^{(a,b)}(u)]$ is symmetric and positive definite, $F(u)$ and $G(u)$ are monotone increasing, on account of (47.23). Further, by (47.24) and the Schwarz inequality, we obtain

$$| H(u_2) - H(u_1) | \leq [(F(u_2) - F(u_1))(G(u_2) - G(u_1))]^{1/2}$$

Hence, by making use of the Schwarz inequality, we obtain

$$(47.25) \qquad \sum_j | H(u_{j+1}) - H(u_j) |$$

$$\leq \left\{ \left(\sum_j [F(u_{j+1}) - F(u_j)] \right)\left(\sum_j [G(u_{j+1}) - G(u_j)] \right) \right\}^{1/2}$$

$$\leq \{[F(\infty) - F(-\infty)][G(\infty) - G(-\infty)]\}^{1/2}$$

$$= \{(f,f)(g,g)\}^{1/2}$$

which indicate that $H(u)$ is of bounded variation over $(-\infty, \infty)$ and that its total variation is less than $\{(f,f)(g,g)\}^{1/2}$, independent of a, b. By partial integration, we obtain, further,

$$(g(x), y_j(x,u)) = (g(x), u^{-1}L_x y_j(x,u))$$
$$= u^{-1}(L_x g(x), y_j(x,u))$$

Hence, we see, by (47.23), that, for $N > 0$,

$$\left\{ \int_{-\infty}^{-N} + \int_{N}^{\infty} \right\} du \left\{ \int_0^u \sum_{j,k=1}^{2} g_j(u)d\rho_{jk}^{(a,b)}(u)g_k(u) \right\}$$

$$\leq N^{-2} \left\{ \int_{-\infty}^{-N} + \int_{N}^{\infty} \right\} du \left\{ \int_0^u \sum_{j,k=1}^{2} (L_x g(x), y_j(x,u))d\rho_{jk}^{(a,b)}(u)(L_s g(s), y_k(s,u)) \right\}$$

$$\leq N^{-2} \int_{-\infty}^{\infty} du \left\{ \int_0^u \sum_{j,k=1}^{2} (L_x g(x), y_j(x,u))d\rho_{jk}^{(a,b)}(u)(L_s g(s), y_k(s,u)) \right\}$$

$$= N^{-2}(L_x g(x), L_x g(x))$$

Now consider the sequences $\{a_{n'}\}$ and $\{b_{n'}\}$, and denote by $H_{n'}(u)$

the function $H(u)$ corresponding to the interval $[a_{n'}, b_{n'}]$. Then, by making use of (47.25), we obtain from the above results that $\{H_{n'}(u)\}$ satisfies the conditions stated in (46.9′). Therefore, letting $n' \to \infty$, we obtain, by Theorem 46.2 and (47.18),

$$(g, f) = \int_{-\infty}^{\infty} du \left\{ \sum_{j,k=1}^{2} \int_{0}^{u} g_j(u) d\rho_{jk}(u) f_k(u) \right\}$$

To sum up, the general expansion theorem reads as follows: For real-valued functions $g(x)$, $f(x)$ which vanish identically for sufficiently large $|x|$ and are twice continuously differentiable over $(-\infty, \infty)$, we have

$$(47.26) \quad \int_{-\infty}^{\infty} g(x) f(x) dx$$
$$= \int_{-\infty}^{\infty} du \left\{ \sum_{j,k=1}^{2} \int_{0}^{u} \left[\int_{-\infty}^{\infty} g(x) y_j(x, u) dx \right] d\rho_{jk}(u) \left[\int_{-\infty}^{\infty} f(s) y_k(s, u) ds \right] \right\}$$

We can prove that (47.26) is valid, in a certain sense, for square integrable functions in the sense of the Lebesgue integral. But the notion of the Lebesgue integral is beyond the scope of this book, so will be omitted.

48. Density matrix

According to (47.15) and (47.18), the density matrix is given by

$$\rho_{11}(u_2) - \rho_{11}(u_1) = \lim_{v \downarrow 0} \frac{1}{\pi} \int_{u_1}^{u_2} \mathscr{I} \frac{1}{m_1(u + iv) - m_2(u + iv)} du$$

$$\rho_{12}(u_2) - \rho_{12}(u_1) = \rho_{21}(u_2) - \rho_{21}(u_1)$$

$$(48.1) \qquad = \lim_{v \downarrow 0} \frac{1}{\pi} \int_{u_1}^{u_2} \mathscr{I} \frac{m_j(u + iv)}{m_1(u + iv) - m_2(u + iv)} du \quad (i = 1, 2)$$

$$\rho_{22}(u_2) - \rho_{22}(u_1) = \lim_{v \downarrow 0} \frac{1}{\pi} \int_{u_1}^{u_2} \mathscr{I} \frac{m_1(u + iv) m_2(u + iv)}{m_1(u + iv) - m_2(u + iv)} du$$

Here $m_1(\lambda)$, $m_2(\lambda)$ are given by (44.8) or (44.9), (44.6) or (44.7), respectively, depending on whether the boundary points $-\infty$, ∞ are in the limit point case or the limit circle case. The quantities $l_a(\lambda)$ in (44.8), (44.9), and $l_b(\lambda)$ in (44.6), (44.7) are given by (44.1) and (43.2) respectively. In the case when the boundary points are both in the limit point case, we may set $\alpha = 0$, $\beta = 0$, and obtain

$$(48.2) \qquad \begin{aligned} m_1(\lambda) &= \lim_{a \to -\infty} \frac{-y_1(a, \lambda)}{y_2(a, \lambda)} \\ m_2(\lambda) &= \lim_{b \to \infty} \frac{-y_1(b, \lambda)}{y_2(b, \lambda)} \end{aligned}$$

However, in the limit circle case, we can not find a simple expression as (48.2). In fact, if, for example, $x = \infty$ is in the limit circle case, the subsequence $\{a_n\}$ in (44.9) is not uniquely determined but depends on α.

The following theorems are useful in practice to calculate the density matrix.

Theorem 48.1. Let $x = \infty$ be in the limit point case, and let $y(x)$ be a solution of

$$y'' + \{\lambda - q(x)\}y = 0$$

such that $y(x) \not\equiv 0$ and $\int_0^\infty |y(x)|^2 dx < \infty$. Then

(48.3) $$m_2(\lambda) = \frac{y'(0)}{y(0)}$$

Proof. According to Theorem 43.4, the function $y(x)$ is a multiple of

$$y_1(x, \lambda) + m_2(\lambda)y_2(x, \lambda)$$

Hence, from

$$y_1(0, \lambda) = 1 , \quad y_1'(0, \lambda) = 0 ; \quad y_2(0, \lambda) = 0 , \quad y_2'(0, \lambda) = 1$$

we obtain (48.3), q.e.d.

THEOREM 48.2. If $q(x)$ is an even function, that is, $q(x) \equiv q(-x)$, then

(48.4) $$m_1(\lambda) \equiv -m_2(\lambda)$$

Accordingly, we have

$$\frac{m_2(\lambda)}{m_1(\lambda) - m_2(\lambda)} = \frac{-1}{2}$$

hence

$$f_{12}(\lambda) = f_{21}(\lambda) = 0$$

so that

(48.5) $$\rho_{12}(u) \equiv \rho_{21}(u) \equiv \text{constant}$$

Proof. If $y(x)$ is a solution of

$$y'' + \{\lambda - q(x)\}y = 0$$

then, so is $y(-x)$. Hence $y_1(-x)$ is a solution of the equation and satisfies the initial conditions

$$y_1(-0) = y_1(0) = 1 , \quad y_1'(-0) = y_1'(0) = 0$$

as well as $y_1(x)$. Therefore, by the uniqueness of the solution, $y_1(x) \equiv y_1(-x)$, hence $y_1(x)$ is an even function. Similarly, we see that $y_2(x)$ is an odd function. Thus, setting

$$a = -b , \quad \alpha = \beta$$

we obtain $l_a(\lambda) \equiv -l_b(\lambda)$, so that $m_1(\lambda) \equiv -m_2(\lambda)$.

THEOREM 48.3. If there exists

$$m_2(u) = \lim_{v \downarrow 0} m_2(u + iv)$$

so that $m_2(u)$ is a real-valued function, then, by (48.1),

(48.6)
$$d\rho_{12}(u) = d\rho_{21}(u) = m_2(u)d\rho_{11}(u)$$
$$d\rho_{22}(u) = m_2(u)^2 d\rho_{11}(u)$$

Accordingly, the expansion theorem becomes

(48.7) $\displaystyle\int_{-\infty}^{\infty} g(x)f(x)dx$

$$= \int_{-\infty}^{\infty} du \left[\int_0^u \left\{ \int_{-\infty}^{\infty} g(x)[y_1(x, u) + m_2(u)y_2(x, u)]dx \right\} \right.$$
$$\left. \times d\rho_{11}(u) \left\{ \int_{-\infty}^{\infty} f(s)[y_1(s, u) + m_2(u)y_2(s, u)]ds \right\} \right]$$

§ 3. Examples

49. *The Fourier series expansion*

We consider the simplest case, with $q(x) \equiv 0$,

(49.1) $y'' + \lambda y = 0 , \quad -1 < x < 1$

Both the boundary points $x = \pm 1$ are in the limit circle case, but non-singular.

Since

(49.2) $y_1(x, \lambda) = \cos(\sqrt{\lambda} \cdot x) , \quad y_2(x, \lambda) = (1/\sqrt{\lambda}) \sin(\sqrt{\lambda} \cdot x)$

we obtain

(49.3)
$$l_a(\lambda) = -\frac{\cos(\sqrt{\lambda} \cdot a) \cos \alpha - \sqrt{\lambda} \cdot \sin(\sqrt{\lambda} \cdot a) \sin \alpha}{1/\sqrt{\lambda} \cdot \sin(\sqrt{\lambda} \cdot a) \cos \alpha + \cos(\sqrt{\lambda} \cdot a) \sin \alpha}$$
$$l_b(\lambda) = -\frac{\cos(\sqrt{\lambda} \cdot b) \cos \beta - \sqrt{\lambda} \cdot \sin(\sqrt{\lambda} \cdot b) \sin \beta}{1/\sqrt{\lambda} \cdot \sin(\sqrt{\lambda} \cdot b) \cos \beta + \cos(\sqrt{\lambda} \cdot b) \sin \beta}$$

In the case

(49.4) $$\alpha = \beta = 0$$

we obtain, by Theorem 48.2,

(49.5) $$m_1(\lambda) = \lim_{a \to -1} \frac{-\sqrt{\lambda} \cdot \cos(\sqrt{\lambda} \cdot a)}{\sin(\sqrt{\lambda} \cdot a)} = \sqrt{\lambda}\,\frac{\cos\sqrt{\lambda}}{\sin\sqrt{\lambda}}$$

$$m_2(\lambda) \equiv -m_2(\lambda)$$

Accordingly, the residues of

$$\frac{-1}{m_1(\lambda) - m_2(\lambda)}, \quad \frac{-m_2(\lambda)}{m_1(\lambda) - m_2(\lambda)}, \quad \frac{-m_1(\lambda)m_2(\lambda)}{m_1(\lambda) - m_2(\lambda)}$$

are

1   at $\sqrt{u} = \frac{1}{2}(2n + 1)\pi$, $n = 0, 1, 2, \cdots$,
0   at every point,
$(n\pi)^2$ at $\sqrt{u} = n\pi$, $n = 1, 2, \cdots$

respectively. Therefore, we see that

$\rho_{11}(u)$ is a monotone increasing step function having discontinuities at the points $(\frac{1}{2}(2n + 1)\pi)^2$ with the jump 1,

$\rho_{12}(u) \equiv \rho_{21}(u) \equiv$ constant,

$\rho_{22}(u)$ is a monotone increasing step function having discontinuities at the points $(n\pi)^2$ with the jump $(n\pi)^2$.

Thus, the general expansion theorem implies

(49.6) $$\int_{-1}^{1} g(x)f(x)dx$$

$$= \sum_{n=0}^{\infty} \int_{-1}^{1} g(x)\cos\left[(2n+1)\frac{\pi}{2}x\right]dx \int_{-1}^{1} f(s)\cos\left[(2n+1)\frac{\pi}{2}s\right]ds$$

$$+ \sum_{n=1}^{\infty} \int_{-1}^{1} g(x)\sin(n\pi x)ds \int_{-1}^{1} f(s)\sin(n\pi s)ds$$

In the case

(49.7) $$\alpha = \beta = \frac{1}{2}\pi$$

we obtain similarly

(49.8) $$\int_{-1}^{1} g(x)f(x)dx$$

$$= \sum_{n=0}^{\infty} \int_{-1}^{1} g(x)\cos(n\pi x)dx \int_{-1}^{1} f(s)\cos(n\pi s)ds$$

$$+ \sum_{n=0}^{\infty} \int_{-1}^{1} g(x)\sin\left[(2n+1)\frac{\pi}{2}x\right]dx \int_{-1}^{1} f(s)\sin\left[(2n+1)\frac{\pi}{2}s\right]ds$$

(49.6) and (49.8) are the Parseval forms of the *Fourier series expansion*.

In the cases when $\alpha = \frac{1}{2}\pi$, $\beta = 0$, or $\alpha = 0$, $\beta = \frac{1}{2}\pi$, or arbitrary α, β, we can obtain the corresponding expansion formulas in a similar manner.

According to the general expansion theorem, these formulas hold for real-valued functions $f(x)$ and $g(x)$ which are twice continuously differentiable on $(-1, 1)$ and identically zero for x sufficiently near the boundary points $-1, 1$. However, we can prove that these formulas hold for every pair of continuous functions $f(x)$, $g(x)$ on $[-1, 1]$; for example, for (49.6), this will be proved as follows.

Let $h(x)$ be a real-valued continuous function on $[-1, 1]$. Then we can choose continuously differentiable real-valued functions $\{f_k(x)\}$ on $(-1, 1)$ which satisfies the following: Every $f_k(x)$ is identically zero for x sufficiently near the boundary points $-1, 1$ and $|h(x)| \geqq |f_k(x)|$, and further $\lim_{k\to\infty} f_k(x) = h(x)$ for every x, $-1 < x < 1$. This is of course possible.

The Fourier coefficient of $f_k(x)$ is the sum of the Fourier coefficient of $h(x)$ and that of $f_k(x) - h(x)$. Hence, by applying the Bessel inequality (21.6) to $h(x)$ and $f_k(x) - h(x)$, we see that the series obtained by setting $f(x) = g(x) = f_k(x)$ in (49.6) is uniformly convergent with respect to k. Therefore, as $k \to \infty$, we can change the order of lim and $\sum$. Hence we obtain that (49.6) holds for $f(x) = g(x) = h(x)$, that is, we have, for every continuous function $h(x)$,

$$\int_{-1}^{1} |h(x)|^2 dx = \sum_{n=0}^{\infty} \left(\int_{-1}^{1} h(x) \cos\left[(2n+1)\frac{\pi}{2}x \right] dx \right)^2$$
$$+ \sum_{n=1}^{\infty} \left(\int_{-1}^{1} h(x) \sin(n\pi x) dx \right)^2$$

Substituting $f(x) + \mu g(x)$ for $h(x)$, we see, by a comparison of the coefficients of μ, that (49.6) holds for all continuous functions $f(x)$ and $g(x)$

50. *The Fourier integral theorem*

We consider the equation

(50.1) $y'' + \lambda y = 0$ $(-\infty < x < +\infty)$

for which we have

(50.2) $y_1(x, \lambda) = \cos(\sqrt{\lambda} \cdot x)$, $y_2(x, \lambda) = \frac{1}{\sqrt{\lambda}} \sin(\sqrt{\lambda} \cdot x)$

The boundary point $x = \infty$ is in the limit point case, since the solution

(50.3) $$z(x, \lambda) = e^{-i x \sqrt{\lambda}}$$

satiesfies

$$\int_{-0}^{\infty} |z(x, \lambda)|^2 dx = \infty$$

for $\mathscr{I}(\lambda) = v > 0$. The point $x = -\infty$ is also in the limit point case. Hence, from (48.2), it follows that, for $\mathscr{I}(\lambda) > 0$,

(50.4)
$$m_1(\lambda) = \lim_{a \to -\infty} \frac{-y_1(a, \lambda)}{y_2(a, \lambda)}$$

$$= -\sqrt{\lambda} \lim_{a \to -\infty} \frac{\cos(\sqrt{\lambda} \cdot a)}{\sin(\sqrt{\lambda} \cdot a)} = -\sqrt{\lambda} \cdot i$$

Furthermore, Theorem 48.2 implies

(50.5) $$m_2(\lambda) \equiv -m_1(\lambda) \equiv \sqrt{\lambda} \cdot i$$

Incidentally, we remark that we can obtain (50.4) also by Theorem 48.1. In fact, let $u(x, \lambda) = e^{i \sqrt{\lambda} x}$. Then $u(0, \lambda) = 1$ and

$$\int_0^{\infty} |u(x, \lambda)|^2 dx < \infty$$

for $\mathscr{I}(\lambda) > 0$. Hence

$$\frac{u'(0, \lambda)}{u(0, \lambda)} = \sqrt{\lambda} \cdot i$$

We thus obtain

$$\pi \frac{d\rho_{11}(u)}{du} = \mathscr{I} \frac{1}{m_1(u) - m_2(u)} = -\mathscr{I} \frac{1}{2\sqrt{u} \cdot i} = \begin{cases} 1/2\sqrt{u}, & 0 < u \\ 0, & u < 0 \end{cases}$$

$$\pi \frac{d\rho_{12}(u)}{du} = \pi \frac{d\rho_{21}(u)}{du} = 0$$

$$\pi \frac{d\rho_{22}(u)}{du} = \mathscr{I} \frac{m_1(u) m_2(u)}{m_1(u) - m_2(u)} = \mathscr{I} \frac{\sqrt{u} \cdot i}{2} = \begin{cases} \sqrt{u}/2, & 0 < u \\ 0, & u < 0 \end{cases}$$

Hence, by the general expansion theorem, we obtain

(50.6) $$\int_{-\infty}^{\infty} g(x) f(x) dx$$

$$= \frac{1}{\pi} \int_0^{\infty} \frac{du}{2\sqrt{u}} \left\{ \left[\int_{-\infty}^{\infty} g(x) \cos(\sqrt{u} \cdot x) dx \int_{-\infty}^{\infty} f(s) \cos(\sqrt{u} \cdot s) ds \right] \right.$$

$$\left. + \left[\int_{-\infty}^{\infty} g(x) \sin(\sqrt{u} \cdot x) dx \int_{-\infty}^{\infty} f(s) \sin(\sqrt{u} \cdot s) ds \right] \right\}$$

(50.6) is the Parseval form of the *Fourier integral formula.*

51.   *The Hermite function expansion*

We consider the equation

(51.1) $y'' + (\lambda - x^2)y = 0$ $(-\infty < x < \infty)$

Putting

(51.2) $y = e^{-\frac{1}{2}x^2}z$

in (51.1), we obtain

(51.3) $z'' - 2xz' + (\lambda - 1)z = 0$

As was shown in Part 24, (51.3) possesses a solution of the form

(51.4) $z(x) = \sum_{n=0}^{\infty} a_{2n}x^{2n}$, $\dfrac{a_{n+2}}{a_n} = \dfrac{(2n - \lambda + 1)}{(n + 1)(n + 2)}$

Let b_n be the coefficient of x^n in the expansion

$$e^{x^2} = 1 + x^2 + \frac{x^4}{2!} + \cdots + \frac{x^n}{(\frac{1}{2}n)!} + \frac{x^{n+2}}{(\frac{1}{2}n + 1)!} + \cdots$$

If $\mathscr{F}(\lambda) = 0$ and λ is not an odd number, then, as $n \to \infty$, $a_n = a_n(\lambda)$ becomes zero of the same order as b_n. In fact, from

$$\frac{b_{n+2}}{b_n} = \frac{1}{\frac{1}{2}n + 1}$$

there follows that

$$\frac{a_{n+2}}{b_{n+2}} \approx \frac{a_n}{b_n} \qquad (n \to \infty)$$

Consequently, if $\mathscr{F}(\lambda) = 0$ and λ is not odd, the function (51.2) with a_0 being 1 becomes infinite of the same order as $e^{\frac{1}{2}x^2}$. Hence, by Theorem 43.4, the point $x = \infty$ is in the limit point case. Similarly, the point $x = -\infty$ is in the limit point case. Further, by Theorem 48.2, we obtain $m_1(\lambda) \equiv -m_2(\lambda)$.

Next, we shall calculate $m_2(\lambda)$ by Theorem 48.1. Consider

(51.5) $z(x) = \displaystyle\int_{\infty}^{(0+)} e^{-x\zeta - \frac{1}{4}\zeta^2} \cdot \zeta^{-\frac{1}{2}\lambda - \frac{1}{2}} d\zeta$

where $\displaystyle\int_{\infty}^{(0+)}$ means complex integration along the contour C in Fig. 3, and the branch of

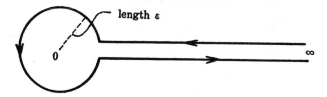

Fig. 3.

$$\zeta^{-\frac{1}{2}\lambda-\frac{1}{2}} = \exp\left\{\left(-\frac{\lambda}{2} - \frac{1}{2}\right)\log\zeta\right\}$$

is taken so that $\log\zeta$ is real on the real line indicated by $\leftarrow$ in Fig. 3. Since

$$z'' - 2xz' + (\lambda - 1)z = \int_{\infty}^{(0+)} e^{-x\zeta-\frac{1}{4}\zeta^2}\cdot\zeta^{-\frac{1}{2}\lambda-\frac{1}{2}}(\zeta^2 + 2x\zeta + \lambda - 1)d\zeta$$

$$= -2^{-1}\int_{\infty}^{(0+)} \frac{d}{d\zeta}(e^{-x\zeta-\frac{1}{4}\zeta^2}\cdot\zeta^{-\frac{1}{2}\lambda+\frac{1}{2}})d\zeta = 0$$

we have that (51.5) is a solution of (51.3), and hence

(51.6) $$y(x, \lambda) = e^{-\frac{1}{2}x^2}\int_{\infty}^{(0+)} e^{-x\zeta-\frac{1}{4}\zeta^2}\cdot\zeta^{-\frac{1}{2}\lambda-\frac{1}{2}}d\zeta$$

is a solution of (51.1). Moreover, it is easily proved by (51.6) that, for fixed λ,

$$|y(x, \lambda)| = O(e^{-\frac{1}{2}x^2+x})$$

as $x \to \infty$. Hence we obtain

$$\int_0^\infty |y(x, \lambda)|^2 dx < \infty$$

Therefore, Theorem 48.1 implies

(51.7) $$m_2(\lambda) = \frac{y'(0, \lambda)}{y(0, \lambda)}$$

$$= -\int_{\infty}^{(0+)} e^{-\frac{1}{4}\zeta^2}\cdot\zeta^{-\frac{1}{2}\lambda+\frac{1}{2}}d\zeta \Big/ \int_{\infty}^{(0+)} e^{-\frac{1}{4}\zeta^2}\cdot\zeta^{-\frac{1}{2}\lambda-\frac{1}{2}}d\zeta$$

On the other hand, when ζ winds once around the origin O in the positive sense,

$$\zeta^{-\frac{1}{2}\lambda\pm\frac{1}{2}}$$

becomes

$$e^{2\pi l(-\frac{1}{2}\lambda\pm\frac{1}{2})} \cdot \zeta^{-\frac{1}{2}\lambda\pm\frac{1}{2}}$$

Therefore, letting $\varepsilon \to 0$ (in Fig. 3), we obtain

(51.8) $\qquad m_2(\lambda) = -\int_0^\infty e^{-\frac{1}{4}\zeta^2} \cdot \zeta^{-\frac{1}{2}\lambda+\frac{1}{2}} d\zeta \Big/ \int_0^\infty e^{-\frac{1}{4}\zeta^2} \cdot \zeta^{-\frac{1}{2}\lambda-\frac{1}{2}} d\zeta$

Setting $\zeta^2 = 4t$ in

(51.9) $\qquad\qquad \Gamma(\lambda) = \int_0^\infty e^{-t} t^{\lambda-1} dt$

we obtain

(51.10) $\qquad\qquad m_2(\lambda) = -2\Gamma(\frac{3}{4} - \frac{1}{4}\lambda)/\Gamma(\frac{1}{4} - \frac{1}{4}\lambda)$

According to Theorem 48.2, $m_1(\lambda) \equiv -m_2(\lambda)$, and we obtain

$$\mathscr{S} \frac{1}{m_1(\lambda) - m_2(\lambda)} = \mathscr{S} \frac{\Gamma(\frac{1}{4} - \frac{1}{4}\lambda)}{4\Gamma(\frac{3}{4} - \frac{1}{4}\lambda)}$$

$$\mathscr{S} \frac{m_2(\lambda)}{m_1(\lambda) - m_2(\lambda)} = 0$$

$$\mathscr{S} \frac{m_1(\lambda)m_2(\lambda)}{m_1(\lambda) - m_2(\lambda)} = -\mathscr{S} \frac{\Gamma(\frac{3}{4} - \frac{1}{4}\lambda)}{\Gamma(\frac{1}{4} - \frac{1}{4}\lambda)}$$

$\Gamma(\lambda)$ is a meromorphic function with simple poles at $\lambda = 0, -1, -2,$ $\cdots$ and without zeros. Therefore, by a similar calculation of residues as in (45.14), we see that

(51.11)
$\rho_{11}(u)$ is a monotone increasing step function having discontinuities at the points $4n + 1$ $(n = 0, 1, 2, \cdots)$,

$\rho_{12}(u) \equiv \rho_{21}(u) \equiv$ constant,

$\rho_{22}(u)$ is a monotone increasing step function having discontinuities at the points $4n + 3$ $(n = 0, 1, 2, \cdots)$.

From Part 24, we obtain

(51.12)
$$y_1(x, 4n + 1) = e^{-\frac{1}{2}x^2} H_{2n}(x)/H_{2n}(0)$$
$$y_2(x, 4n + 3) = e^{-\frac{1}{2}x^2} H_{2n+1}(x)/H'_{2n+1}(0)$$

For, if we replace λ by $\lambda - 1$ in (24.3), the equation (24.3) becomes (51.3), and hence $y_1(x, \lambda)$ and $y_2(x, \lambda)$ are given by (51.2) and the power series in (24.4') with $a_0 = 1$ and $a_1 = 1$, respectively.

Thus, we obtain, by the general expansion theorem,

(51.13) $\qquad \int_{-\infty}^\infty g(x) f(x) dx$

$$= \sum_{n=0}^\infty C_n \int_{-\infty}^\infty e^{-\frac{1}{2}x^2} g(x) H_n(x) dx \int_{-\infty}^\infty e^{-\frac{1}{2}s^2} f(s) H_n(s) ds$$

where C_n is a positive number, independent of $f(x)$ and $g(x)$. To determine C_n, we use the following: According to the general expansion theorem, (51.13) holds for functions $f(x)$ and $g(x)$ which are twice continuously differentiable and vanish identically for sufficiently large $|x|$, but it also holds for

$$f(x) = g(x) = e^{-\frac{1}{2}x^2}H_n(x)$$

which can be proved similarly as at the end of Part 49. Substitute these in (51.13). Then the left side becomes $2^n\sqrt{\pi}\cdot(n!)$, because of (24.14). On the other hand, the right side becomes, by the orthogonality relations (24.8),

$$C_n\left\{\int_{-\infty}^{\infty} e^{-x^2}H_n(x)^2dx\right\}^2 = C_n[2^n\sqrt{\pi}\cdot(n!)]^2$$

Therefore, we obtain

(51.14)
$$C_n = \frac{1}{2^n(n!)\sqrt{\pi}}$$

Accordingly, we have the Parseval form (51.13)–(51.14) of the expansion in the Hermite functions $\{e^{-\frac{1}{2}x^2}H_n(x)\}$. This also shows the completeness of the Hermite functions. (Compare with the completeness of the Legendre polynomials in Part 24.) Incidentally, the completeness of the trigonometric functions is also shown in Part 49, in this sense.

52. *The Hankel integral theorem*

We consider the equation

(52.1)
$$\frac{d^2y}{dx^2} + \left\{\lambda - \frac{\nu^2 - \frac{1}{4}}{x^2}\right\}y = 0 \qquad (0 < x < \infty)$$

Putting

(52.2)
$$y = x^{\frac{1}{2}}z$$

we obtain

(52.3)
$$x^2z'' + xz' + (\lambda x^2 - \nu^2)z = 0$$

Hence, as was shown in Part 15, a fundamental system of solutions of (52.1) is given by

$$x^{\frac{1}{2}}J_\nu(\sqrt{\lambda}\cdot x), \quad x^{\frac{1}{2}}Y_\nu(\sqrt{\lambda}\cdot x)$$

Let c be an arbitrary number such that $0 < c < \infty$. Then, by making use of (15.10) and Lommel's formula (15.14), the solutions satisfying the initial conditions

(52.4)
$$y_1(c, \lambda) = 1 , \quad y_1'(c, \lambda) = 0$$
$$y_2(c, \lambda) = 0 , \quad y_2'(c, \lambda) = 1$$

are given by

$$y_1(x, \lambda) = \frac{\pi}{2} x^{\frac{1}{2}} c^{\frac{1}{2}} \sqrt{\lambda} \{ J_\nu(\sqrt{\lambda} \cdot x) Y_\nu'(\sqrt{\lambda} \cdot c)$$

(52.5)
$$- Y_\nu(\sqrt{\lambda} \cdot x) J_\nu'(\sqrt{\lambda} \cdot c) \} - \frac{1}{2c} y_2(x, \lambda)$$

$$y_2(x, \lambda) = \frac{-\pi}{2} x^{\frac{1}{2}} c^{\frac{1}{2}} \{ J_\nu(\sqrt{\lambda} \cdot x) Y_\nu(\sqrt{\lambda} \cdot c)$$

$$- Y_\nu(\sqrt{\lambda} \cdot x) J_\nu(\sqrt{\lambda} \cdot c) \}$$

In the following, we shall restrict ourselves to the case when $\nu \geqq 1$. From (15.7), it follows that

$$\int_0^c | x^{\frac{1}{2}} J_{-\nu}(\sqrt{\lambda} \cdot x) |^2 dx = \infty$$

for the integrand tends to infinity as $x \to 0$, in such a way that the integral diverges. Hence the boundary point $x = 0$ is in the limit point case.

From (15.6), it also follows that

$$\int_0^c | x^{\frac{1}{2}} J_\nu(\sqrt{\lambda} \cdot x) |^2 dx < \infty$$

Hence, by Remark 2 to Theorem 43.4, the expression $y_1(x,\lambda) + m_1(\lambda) y_2(x,\lambda)$ is a multiple of $x^{\frac{1}{2}} J_\nu(\sqrt{\lambda} \cdot x)$, that is, by (52.4)

(52.6)
$$y_1(x, \lambda) + m_1(\lambda) y_2(x, \lambda) = \frac{x^{\frac{1}{2}} J_\nu(\sqrt{\lambda} \cdot x)}{c^{\frac{1}{2}} J_\nu(\sqrt{\lambda} \cdot c)}$$

Therefore, as in Theorem 48.1, we obtain, by (52.4),

(52.7)
$$m_1(\lambda) = \frac{\sqrt{\lambda} \, J_\nu'(\sqrt{\lambda} \cdot c)}{J_\nu(\sqrt{\lambda} \cdot c)} + \frac{1}{2c}$$

From (15.21), it follows that

$$\int_c^\infty | x^{\frac{1}{2}} J_n(x) |^2 dx = \infty$$

Hence, the boundary point $x = \infty$ is also in the limit point case. We use the following formula[1]: as $|z| \to \infty$

$$(52.8) \qquad H_\nu^{(1)}(z) = \left(\frac{2}{\pi z}\right)^{\frac{1}{2}} \frac{\exp\left[i(z - \nu\frac{1}{2}\pi - \frac{1}{4}\pi)\right]}{\varGamma(\nu + \frac{1}{2})}$$

$$\times \left[\sum_{\mu=0}^{p-1} \binom{\nu - \frac{1}{2}}{\mu} \varGamma(\nu + \mu + \frac{1}{2})\left(\frac{i}{2z}\right)^\mu + O(|z|^{-p})\right]$$

where $\mathscr{R}(\nu - \frac{1}{2} - p) < 0$. This implies that

$$(52.9) \qquad \int_c^\infty |x^{\frac{1}{2}} H_\nu^{(1)}(\sqrt{\lambda}\cdot x)|^2 dx < \infty$$

if $\mathscr{I}(\lambda) > 0$. Hence, by Theorem 43.4, $y_1(x, \lambda) + m_2(\lambda)y_2(x, \lambda)$ is a multiple of $x^{\frac{1}{2}} H_\nu^{(1)}(\sqrt{\lambda}\cdot x)$, that is, by (52.4)

$$(52.10) \qquad y_1(x, \lambda) + m_2(\lambda)y_2(x, \lambda) = \frac{x^{\frac{1}{2}} H_\nu^{(1)}(\sqrt{\lambda}\cdot x)}{c^{\frac{1}{2}} H_\nu^{(1)}(\sqrt{\lambda}\cdot c)}$$

Therefore, by Theorem 48.1, we obtain

$$(52.11) \qquad m_2(\lambda) = \frac{\sqrt{\lambda}\, H_\nu^{(1)\prime}(\sqrt{\lambda}\cdot c)}{H_\nu^{(1)}(\sqrt{\lambda}\cdot c)} + \frac{1}{2c}$$

By making use of Lommel's formula (15.14) and the definition of $H_\nu^{(1)}$, we obtain further

$$\mathscr{I}\,\frac{1}{m_1(\lambda) - m_2(\lambda)} = -\,\mathscr{I}\left\{\frac{\pi c}{2i} J_\nu(\sqrt{\lambda}\cdot c)H_\nu^{(1)}(\sqrt{\lambda}\cdot c)\right\}$$

$$= \frac{\pi c}{2} J_\nu^2(\sqrt{\lambda}\cdot c) \qquad\qquad (\lambda > 0)$$

since $J_\nu(x)$ and $Y_\nu(x)$ are real for real x. Furthermore, from (15.6) and (15.16), it follows at once that

$$-\mathscr{I}\left\{\frac{\pi c}{2i} J_\nu(\sqrt{\lambda}\cdot c)H_\nu^{(1)}(\sqrt{\lambda}\cdot c)\right\} = 0 \qquad\qquad (\lambda < 0)$$

Accordingly, putting $s^2 = \lambda$, we obtain, by Theorem 48.3, that

$$(52.12) \qquad \int_{-\infty}^\infty g(x)f(x)dx$$

$$= \frac{1}{\pi}\int_0^\infty du\left\{\int_0^u du\left[\int_0^\infty g(x)x^{\frac{1}{2}} J_\nu(\sqrt{u}\cdot x)dx \int_0^\infty f(s)s^{\frac{1}{2}} J_\nu(\sqrt{u}\cdot s)ds\right]\right\}$$

[1] R. Courant and D. Hilbert, *Methods of Mathematical Physics*, Vol. 1, Interscience, New York, 1953, p. 525.

This is the Parseval form of the *Hankel integral theorem*.

In general, we can prove (52.12) for all $\nu \geq 0$. For $\nu = \frac{1}{2}$ the equation (52.12) becomes the Fourier sine integral theorem, which is easily derived from (15.17) and (15.18). The proof will be omitted. Concerning this, the reader should refer to E. C. Titchmarsh, *Eigenfunction Expansions Associated with Second Order Differential Equations*, Oxford, 1946.

53. *The Fourier-Bessel series expansion*

We consider

$$(53.1) \qquad y'' + \left\{\lambda - \frac{\nu^2 - \frac{1}{4}}{x^2}\right\} y = 0 \qquad (0 < x < b < \infty)$$

We shall restrict ourselves to the case when $\nu \geq 1$. Let c be an arbitrary number such that $0 < c < b$. Then, the fundamental system, corresponding to the initial conditions (52.4) at the point c, is given by (52.5). Hence, we obtain (52.6) and (52.7).

The boundary point $x = b$ is of course in the limit circle case. Putting $\beta = 0$ in the boundary condition at $x = b$, we obtain

$$(53.2) \qquad m_2(\lambda) = \lim_{b' \to b} \frac{-y_1(b', \lambda)}{y_2(b', \lambda)} = \frac{-y_1(b, \lambda)}{y_2(b, \lambda)}$$

Hence, we obtain, by (52.6),

$$(53.3) \qquad m_1(\lambda) - m_2(\lambda) = m_1(\lambda) + \frac{y_1(b, \lambda)}{y_2(b, \lambda)}$$

$$= \frac{y_1(b, \lambda) + m_1(\lambda)y_2(b, \lambda)}{y_2(b, \lambda)} = \frac{b^{\frac{1}{2}} J_\nu(\sqrt{\lambda} \cdot b)}{y_2(b, \lambda) c^{\frac{1}{2}} J_\nu(\sqrt{\lambda} \cdot c)}$$

Hence,

$$(53.4) \qquad \frac{-1}{m_1(\lambda) - m_2(\lambda)}$$

is a meromorphic function of λ. Accordingly, by (53.3), we see, similarly as in the calculation of residues in Part 45, that every element of the density matrix is a step function and has discontinuities at the poles of (53.4), which are

$$(53.5) \qquad \text{the zeros of } J_\nu(\sqrt{\lambda} \cdot b), \text{ except for } \lambda = 0.$$

We denote these zeros of $J_\nu(\sqrt{\lambda} \cdot b)$ by $\lambda_1, \lambda_2, \lambda_3, \cdots$. Then every λ_n is positive, since it is known that, for $\nu > -1$, $J_\nu(z)$ has only real zeros.[1] The residue of (53.4) at λ_n is also positive, because the density matrix (47.19) is positive definite.

Put $\lambda = s^2$, so that $\sqrt{\lambda_n} = s_n$. Then, since $m_2(s^2) = -y_1(b, s^2)/y_2(b, s^2)$ is real for real s and $m_1(s_n^2) = m_2(s_n^2)$, Theorem 48.3 implies that

$$
(53.6) \quad \int_0^b g(x)f(x)dx
$$

$$
= \sum_{n=1}^{\infty} c_n' \int_0^b g(x)\{ y_1(x, s_n^2) + m_2(s_n^2)y_2(x, s_n^2)\}dx
$$

$$
\times \int_0^b f(t)\{ y_1(t, s_n^2) + m_2(s_n^2)y_2(t, s_n^2)\}dt
$$

$$
= \sum_{n=1}^{\infty} c_n \int_0^b g(x)x^{\frac{1}{2}}J_\nu(s_nx)dx \int_0^b f(t)t^{\frac{1}{2}}J_\nu(s_nt)dt
$$

where c_n is a positive number independent of $f(x)$ and $g(x)$.

To determine c_n, we first prove the orthogonality relations

$$
(53.7) \quad \int_0^b xJ_\nu(s_nx)J_\nu(s_mx)dx = 0 \qquad (n \neq m)
$$

as follows. From (Part 15)

$$
J_\nu'' + \frac{1}{x}J_\nu' + \left(1 - \frac{\nu^2}{x^2}\right)J_\nu = 0
$$

there follows that

$$
J_\nu''(sx) + \frac{1}{sx}J_\nu'(sx) + \left(1 - \frac{\nu^2}{s^2x^2}\right)J_\nu(sx) = 0
$$

Hence we obtain

$$
x\{s_n^2J_\nu''(s_nx)J_\nu(s_mx) - s_m^2J_\nu''(s_mx)J_\nu(s_nx)\}
$$
$$
+ \{s_nJ_\nu'(s_nx)J_\nu(s_mx) - s_mJ_\nu'(s_mx)J_\nu(s_nx)\}
$$
$$
+ (s_n^2 - s_m^2)xJ_\nu(s_nx)J_\nu(s_mx) = 0
$$

The sum of the first and the second terms is equal to the derivative of

$$
(53.8) \quad x\{s_nJ_\nu'(s_nx)J_\nu(s_mx) - s_mJ_\nu'(s_mx)J_\nu(s_nx)\}
$$

with respect to x. Hence, integrating this from 0 to b and remembering $\nu \geqq 1$ and $J_\nu(s_nb) = J_\nu(s_mb) = 0$, we see that the following equality is true.

[1] R. Courant and D. Hilbert, *Methods of Mathematical Physics*, Vol. 1, Interscience, New York, 1953, p. 494.

$$\int_0^b (s_n^2 - s_m^2) x J_\nu(s_n x) J_\nu(s_m x) dx = 0$$

Since $s_n \neq s_m$ for $n \neq m$, the equation (53.7) is proved.

As in Part 51 it is proved that (53.6) also holds for

$$g(x) = f(x) = x^{\frac{1}{2}} J_\nu(s_n x)$$

Hence, substituting these in (53.6), we obtain

$$(53.9) \qquad c_n^{-1} = \int_0^b x J_\nu(s_n x)^2 dx \qquad (n = 1, 2, \cdots)$$

The integral on the right side is calculated as follows. We can prove, as before, that

$$(53.10) \qquad \int_0^b x J_\nu(\alpha x) J_\nu(\beta x) dx = \frac{b}{\alpha^2 - \beta^2} \{\beta J_\nu(\alpha b) J_\nu'(\beta b) - \alpha J_\nu'(\alpha b) J_\nu(\beta b)\}$$

Put $\alpha = \beta + \varepsilon$ in (53.10) and differentiate the denominator and the numerator on the right side separately with respect to ε. Then, setting $\varepsilon = 0$, we obtain

$$\int_0^b x J_\nu(\beta x)^2 dx = \frac{b^2}{2\beta} \left\{ \beta J_\nu'(\beta b)^2 - J_\nu'(\beta b) J_\nu(\beta b) \frac{1}{b} - \beta J_\nu''(\beta b) J_\nu(\beta b) \right\}$$

Put $\beta = s_n$. Then we have, by $J_\nu(s_n b) = 0$,

$$(53.11) \qquad c_n^{-1} = \int_0^b x J_\nu(s_n x)^2 dx = \frac{b^2}{2} J_\nu'(s_n b)^2$$

Consequently, we obtain the Parseval form of the *Fourier-Bessel series expansion*, in the form

$$(53.12) \qquad \int_0^b g(x) f(x) dx$$

$$= \sum_{n=1}^\infty \frac{2}{b^2 J_\nu'(s_n b)^2} \int_0^b g(x) x^{\frac{1}{2}} J_\nu(s_n x) dx \int_0^b f(t) t^{\frac{1}{2}} J_\nu(s_n t) dt$$

This shows the completeness of the orthogonal system

$$(53.13) \qquad x^{\frac{1}{2}} J_\nu(s_n x) \qquad (n = 1, 2, \cdots)$$

on $[0, b]$. Accordingly, these functions are the eigenfunctions of the singular boundary value problem:

$$y'' + \left\{ s_n^2 - \frac{\nu^2 - \frac{1}{4}}{x^2} \right\} y = 0 \; ;$$

(53.14)
as $x \to 0$, y tends to zero of the same orders as $x^{\nu+\frac{1}{2}}$;
at $x = b$, y vanishes.

In general, we obtain in this way the Fourier-Bessel series expansions for $\nu \geqq 0$, which, for $\nu = \frac{1}{2}$, becomes the Fourier sine series expansion, but we will not enter into the details. For these, reference may be made to Titchmarsh's book mentioned before.

54. *The Laguerre function expansion*

The Laguerre polynomial $L_n(x)$ is a solution of the equation

$$(54.1) \qquad xy'' + (1 - x)y' + \lambda y = 0 \qquad\qquad 0 < x < \infty$$

for $\lambda = n$ (Part 24). Now consider the transformation

$$(54.2) \qquad x = t^2 , \qquad z = t^{\frac{1}{2}}e^{-\frac{1}{2}t^2}y$$

Then we have, for $\lambda = n$, the equation

$$(54.3) \qquad \frac{d^2z}{dt^2} = \left(-4n - 2 + t^2 - \frac{1}{4t^2}\right)z \qquad\qquad (0 < t < \infty)$$

which possesses the solution

$$(54.4) \qquad \sqrt{2}\cdot t^{\frac{1}{2}}e^{-\frac{1}{2}t^2}L_n(t^2) \qquad\qquad (n = 0, 1, 2, \cdots)$$

As was shown in Part 24, the system $\{e^{-\frac{1}{2}x}L_n(x)\}$ is an orthonormal system on $(0, \infty)$; hence the functions in (54.4) also form an orthonormal system on $(0, \infty)$, that is, we have

$$(54.5) \qquad \int_0^\infty 2te^{-t^2}L_n(t^2)L_m(t^2)dt = \delta_{nm}$$

The Parseval relation for (54.5), which follows from the general expansion theorem,

$$(54.6) \qquad \int_0^\infty g(t)f(t)dt$$

$$= \sum_{n=0}^\infty \left\{\int_0^\infty g(t)\sqrt{2}\cdot t^{\frac{1}{2}}e^{-\frac{1}{2}t^2}L_n(t^2)dt\int_0^\infty f(s)\sqrt{2}\cdot s^{\frac{1}{2}}e^{-\frac{1}{2}s^2}L_n(s^2)ds\right\}$$

shows the completeness of the system of the Laguerre functions (54.4), and hence of the system $\{e^{-\frac{1}{2}x}L_n(x)\}$. However, to derive (54.6) from the general expansion theorem, we need a certain function theoretical preparation, which cannot be given in this book. Therefore, we refer the reader to Titchmarsh's book, concerning this derivation. We shall instead give a direct proof due to J. von Neumann, which is based upon the same idea which was used in the proof of the completeness of the Legendre polynomials in Part 24.

Proof of (54.6). By setting $t^2 = x$ in (54.6), we obtain

(54.6′) $$\int_0^\infty g(x)f(x)dx = \sum_{n=0}^\infty \left\{ \int_0^\infty g(x)e^{-\frac{1}{2}x}L_n(x)dx \int_0^\infty f(s)e^{-\frac{1}{2}s}L_n(s)ds \right\}$$

Hence it is sufficient to prove (54.6′) for the functions $f(x)$ and $g(x)$ which have the following property:

(*) $f(x)$ and $g(x)$ are twice continuously differentiable on $(0, \infty)$, and are identically zero near the boundary points $x = 0$ and $x = \infty$.

As in the case of the Legendre polynomials (Part 24), it is sufficient to show the following: for a given $f(x)$ with the property (*), and $\varepsilon > 0$, there exist constants $c_1, c_2, \cdots, c_k$ such that

$$\int_0^\infty \left| f(x) - \sum_{n=0}^k e^{-\frac{1}{2}x}L_n(x)c_n \right|^2 dx < \varepsilon$$

On the other hand, since $f(x)$ has the property (*), it may be written as

(54.7) $$f(x) = e^{-\frac{1}{2}x}h(x)$$

where $h(x)$ has the property (*). Furthermore, $L_n(x)$ is a polynomial of degree n. Therefore, we see that it is sufficient to show the following: for a given $h(x)$ with the property (*), and $\varepsilon > 0$, there exists a polynomial $p_k(x)$ of degree k such that

(54.8) $$\int_0^\infty e^{-x} | h(x) - p_k(x) |^2 dx < \varepsilon$$

Since $h(x)$ has the property (*), $h(\log(1/y))$ is continuous on the interval $0 \leq y \leq 1$. Hence, by the Weierstrass approximation theorem, there exists, for any $\varepsilon' > 0$, a polynomial $\rho(y)$ such that

$$\sup_{0 \leq y \leq 1} \left| h\left(\log \frac{1}{y}\right) - \rho(y) \right| < \varepsilon'$$

Hence

$$\int_0^1 \left| h\left(\log \frac{1}{y}\right) - \rho(y) \right|^2 dy < \varepsilon'$$

so that

(54.9) $$\int_0^\infty e^{-x} | h(x) - \rho(e^{-x}) |^2 dx < \varepsilon'$$

Suppose that, for any integer $m \geq 0$ and $\delta > 0$, there exists a polynomial $p_{m,\delta}(x)$ such that

$$(54.10) \qquad \int_0^\infty e^{-x} |e^{-mx} - p_{m,\delta}(x)|^2 dx < \delta$$

We then use the inequality (20.6), i.e.

$$\|\alpha + \beta\| = \left(\int_0^\infty |\alpha(x) + \beta(x)|^2 dx \right)^{\frac{1}{2}} \leq \|\alpha\| + \|\beta\|$$

which can be proved by virtue of Schwarz' inequality. Thus, setting $\rho(y) = \sum_{n=0}^s c_n y^n$, we obtain

$$\left(\int_0^\infty e^{-x} \left| h(x) - \sum_{n=0}^s c_n p_{n,\delta}(x) \right|^2 dx \right)^{\frac{1}{2}}$$

$$\leq \left(\int_0^\infty e^{-x} |h(x) - \rho(e^{-x})|^2 dx \right)^{\frac{1}{2}}$$

$$+ \sum_{n=0}^s |c_n| \left(\int_0^\infty e^{-x} |e^{-nx} - p_{n,\delta}(x)|^2 dx \right)^{\frac{1}{2}}$$

The second term on the right side tends to zero as $\delta \to 0$ because of (54.10), (s and c_n are independent of δ). Since $\varepsilon' > 0$ was arbitrary, the proof is completed. It remains to prove (54.10).

Proof of (54.10). We use (24.28), that is,

$$(54.11) \qquad \sum_{n=0}^\infty L_n(x) t^n = \frac{1}{1-t} \exp\left(-\frac{xt}{1-t} \right) \qquad (|t| < 1)$$

Set

$$(54.12) \qquad t = \frac{m}{m+1}, \text{ i.e., } m = \frac{t}{1-t}$$

Then we obtain

$$e^{-mx} - (1-t) \sum_{n=0}^N L_n(x) t^n = (1-t) \sum_{n=N+1}^\infty L_n(x) t^n$$

Hence, by virtue of the orthonormality of $\{e^{-\frac{1}{2}x} L_n(x)\}$, we have

$$\int_0^\infty \left| e^{-mx} - (1-t) \sum_{n=0}^N L_n(x) t^n \right|^2 e^{-x} dx$$

$$= (1-t)^2 \int_0^\infty e^{-x} \left| \sum_{n=N+1}^\infty L_n(x) t^n \right|^2 dx = (1-t)^2 \sum_{n=N+1}^\infty t^{2n}$$

The last term tends to zero, as $N \to \infty$. Thus the proof is completed. In the above reasoning, the exchange of the order of $\sum\limits_{n=N+1}^{\infty}$ and $\int_0^{\infty}$ is permissible, since the estimate

$$\sum_{n,n'=N+1}^{\infty}\int_0^{\infty}e^{-x}|L_n(x)|\,|L_{n'}(x)|\,dx\cdot t^{n+n'}$$

$$\leq \sum_{n,n'=N+1}^{\infty}\left\{\int_0^{\infty}e^{-x}L_n(x)^2dx\right\}^{\frac{1}{2}}\left\{\int_0^{\infty}e^{-x}L_{n'}(x)^2dx\right\}^{\frac{1}{2}}t^{n+n'}$$

$$= \sum_{n,n'=N+1}^{\infty}t^{n+n'}$$

shows that the series of positive terms on the left side is convergent for $|t| < 1$.

CHAPTER 6

NON-LINEAR INTEGRAL EQUATIONS

The Volterra integral equation

$$\varphi(x) - \lambda \int_a^x K(x, s)\varphi(s)ds = f(x)$$

and the Fredholm integral equation

$$\varphi(x) - \lambda \int_a^b K(x, s)\varphi(s)ds = f(x)$$

with which we were concerned in the preceding chapters, 2 to 5, are both *linear* with respect to the unknown function φ. This linearity was the cause for the relatively smooth development of the general theory of these equations.

The theory of *non-linear integral equations* is very important in pure and applied mathematics. We have already observed in Chapter 1 that the initial value problem for ordinary differential equations can be reduced to a non-linear Volterra integral equation. However, for the general treatment of non-linear integral equations, we need the knowledge of functional analysis, which is beyond the scope of this book. In this chapter we shall give a brief sketch of several classical results for non-linear integral equations.

55. *Non-linear Volterra integral equations*

We consider the equation

(55.1) $$\varphi(x) + \int_0^x F(x, s, \varphi(s))ds = f(x)$$

in the unknown $\varphi(x)$. We make the following assumptions. The function $F(x, y, z)$ is continuous on a domain D defined by

$$|x| \leqq a, \quad |y| \leqq a, \quad |z| \leqq b$$

and satisfies the Lipschitz condition with respect to z

(55.2) $$|F(x, y, z_1) - F(x, y, z_2)| \leq K|z_1 - z_2|$$

in D.

The function $f(x)$ is continuous for $|x| \leqq a$, vanishes for $x = 0$, and satisfies the Lipschitz condition

(55.3) $$|f(x_1) - f(x_2)| \leqq k|x_1 - x_2|$$

Let D' be a domain given by

(55.4) $$|x| \leqq a', \quad a' = \min\left(a, \frac{b}{k+M}\right), \quad M = \sup_D F(x, y, z)$$

Then we can define on D' the successive approximation functions

$$\varphi_0(x) = f(x)$$
$$\varphi_1(x) = f(x) - \int_0^x F(x, s, \varphi_0(s))ds$$
$$\dots\dots\dots\dots\dots\dots$$
$$\varphi_n(x) = f(x) - \int_0^x F(x, s, \varphi_{n-1}(s))ds$$
$$\dots\dots\dots\dots\dots\dots$$

We can prove as in Part 1 that the $\{\varphi_n(x)\}$ is uniformly convergent on D' and the limit

$$\lim_{n\to\infty}\varphi_n(x) = \varphi(x)$$

is a unique solution of the equation (55.1).

56. *Non-linear Fredholm integral equations*

We consider the equation

(56.1) $$\varphi(x) + \lambda\int_a^b F(x, s, \varphi(s))ds = f(x)$$

in the unknown $\varphi(x)$. We make the following assumptions. The function $F(x, y, z)$ is continuous on a domain D defined by

$$|x| \leqq a, \quad |y| \leqq b, \quad |z| \leqq c$$

and satisfies the Lipschitz condition with respect to z

(56.2) $$|F(x, y, z_1) - F(x, y, z_2)| \leqq K|z_1 - z_2|$$

in D.

The function $f(x)$ is continuous for $a \leqq x \leqq b$ and

(56.3) $$\sup_{a \leqq x \leqq b}|f(x)| = f < c$$

Let

(56.4) $$|\lambda| \leqq \frac{(c-f)}{M(b-a)}, \quad M = \sup_D |F(x, y, z)|$$

Then, for such λ, we can define on $[a, b]$ the successive approximation functions

$$\varphi_0(x) = f(x)$$

$$\varphi_1(x) = f(x) - \lambda \int_a^b F(x, s, \varphi_0(s))ds$$

$$\cdots\cdots\cdots\cdots\cdots$$

$$\varphi_n(x) = f(s) - \lambda \int_a^b F(x, s, \varphi_{n-1}(s))ds$$

$$\cdots\cdots\cdots\cdots\cdots$$

We can prove as in 1 that the sequence $\{\varphi_n(x)\}$ is uniformly convergent on $[a, b]$ and the limit

$$\lim_{n\to\infty}\varphi_n(x) = \varphi(x)$$

is a unique solution of the equation (56.1).

REMARK. It is not necessary that the equation (56.1) contains the parameter λ linearly, but it is necessary that the term containing the integral tends to zero as $\lambda \to 0$. Accordingly, a non-linear equation of the form

$$(56.5) \qquad \varphi(x) + \sum_{m=1}^\infty \lambda^m \int_a^b\int_a^b \cdots \int_a^b K_m(x, s_1, s_2, \cdots, s_m)$$

$$\times F_m[\varphi(s_1), \varphi(s_2), \cdots, \varphi(s_m)]ds_1 ds_2 \cdots ds_m$$

$$= f(x)$$

can also be solved, for sufficiently small $|\lambda|$, by the method of successive approximations.

57. *Schmidt's method*

Schmidt considered non-linear Fredholm integral equations of the form

$$(57.1) \qquad \varphi(x) + \lambda \int_a^b K(x, s)\varphi(s)ds$$

$$+ \sum \lambda^{\alpha_1+\beta_1+\cdots+\alpha_m+\beta_m} \int_a^b \cdots \int_a^b K(x, s_1, \cdots, s_m)$$

$$\times \varphi(s_1)^{\alpha_1}h(s_1)^{\beta_1}\cdots\varphi(s_m)^{\alpha_m}h(s_m)^{\beta_m}ds_1 ds_2 \cdots ds_m$$

$$= f(x)$$

Here $h(s)$ is a given function and $\sum$ means the summation over all the sets $(\alpha_1, \beta_1, \cdots, \alpha_m, \beta_m)$ of non-negative integers with $\alpha_j + \beta_j > 0$.

We write (57.1) as

$$(57.2) \qquad \varphi(x) + \lambda \int_a^b K(x, s)\varphi(s)ds = F(x)$$

In the case when $-\lambda$ is not an eigenvalue of the kernel $K(x, s)$, (57.2), and hence (57.1), is, by Part 28, equivalent to an equation of the form (56.5). Hence, for sufficiently small $|\lambda|$, the equation (57.1) possesses a unique solution. Next we consider the case when $-\lambda$ is an eigenvalue of $K(x, s)$ with multiplicity 1, φ_1 being the corresponding eigenfunction. If there exists a solution φ of the equation (57.2), then, as was shown in Part 28–29, φ can be written in the following form, containing an arbitrary constant C,

$$(57.3) \qquad \varphi(x) = F(x) + \lambda \int_a^b \Gamma_1(x, s)F(s)ds + G(x) + C\varphi_1(x)$$

where $G(x)$, and $\Gamma_1(x, s)$ are known. Furthermore the necessary and sufficient condition for the existence of a solution is that

$$(57.4) \qquad \int_a^b F(s)\varphi_1(s)ds = 0$$

as was shown in Part 29. On the other hand, (57.3) is an equation of the form (56.5), hence it possesses a unique solution for sufficiently small $|\lambda|$ when C is fixed. Therefore, we can find, for sufficiently small $|\lambda|$, the constant C and the solution φ from (57.3) and (57.4).

In general, if $-\lambda$ is an eigenvalue of $K(x, s)$ with multiplicity $m \geq 2$, the problem can be discussed in a similar manner. In such cases, the solution generally contains $m - 1$ arbitrary constants.

APPENDIX

FROM THE THEORY OF FUNCTIONS
OF A COMPLEX VARIABLE

A theorem on normal family of regular functions (Part 44)

Let $\{f_n(z)\}$ be a sequence of regular functions in a domain $|z| < c$. If

$$\sup_{n \geq 1, |z| < c} |f_n(z)| = M < \infty$$

then, for any a, $0 < a < c$, there exists a subsequence $\{f_{n'}(z)\}$ which converges uniformly on the domain $|z| \leq a$.

Proof. As in the case of Ascoli-Arzelà theorem (Part 19), it is sufficient to show that $\{f_n(z)\}$ is equi-continuous for $|z| \leq a$. For any b, $a < b < c$, we obtain, Cauchy's integral formula

$$f_n(z) = \frac{1}{2\pi i} \int_{|\xi| = b} \frac{f_n(\xi)}{\xi - z} d\xi , \qquad |z| \leq a$$

Hence, for any z_1, z_2 such that $|z_1| \leq a$, $|z_2| \leq a$, we have

$$|f_n(z_1) - f_n(z_2)| = \frac{1}{2\pi} \left| \int_{|\xi| = b} f_n(\xi) \left[\frac{1}{\xi - z_1} - \frac{1}{\xi - z_2} \right] d\xi \right|$$

$$\leq \frac{1}{2\pi} \int_{|\xi| = b} |f_n(\xi)| \left| \frac{z_1 - z_2}{(\xi - z_1)(\xi - z_2)} \right| |d\xi|$$

$$\leq \frac{M}{2\pi} |z_1 - z_2| \frac{1}{|b - a|^2} 2\pi b$$

$$= \frac{bM |z_1 - z_2|}{(b - a)^2}$$

This shows that $\{f_n(z)\}$ is equi-continuous, q.e.d.

Hurwitz's theorem (Part 47)

Let $\{f_n(z)\}$ be a sequence of regular functions in a domain D, and let $\lim_{n \to \infty} f_n(z) = f(z)$ uniformly in D. If $f(z)$ is not identically zero and $f(z_0) = 0$ at an inner point z_0 of D, then, for any $\varepsilon > 0$, there exists a number N such that $f_n(z)$ has a zero in a domain $|z - z_0| < \varepsilon$ whenever $n \geq N$.

Proof. According to the assumption, z_0 is an isolated zero of $f(z)$. Hence, for suitably chosen δ ($0 < \delta < \varepsilon$) there exists an $m > 0$ such that

$$|f(z)| \geqq m > 0$$

on the circle $|z - z_0| = \delta$. Furthermore, because of the uniform convergence, we can find N such that

$$|f_n(z) - f(z)| < m$$

on the circle $|z - z_0| = \delta$ whenever $n \geqq N$. From Rouché's theorem, it follows that, for $n \geqq N$,

$$f_n(z) = f(z) + \{f_n(z) - f(z)\}$$

has the same number of zeros in $|z - z_0| < \delta$ as $f(z)$. Since $f(z_0) = 0$, the number be at least one; that is, for $n \geqq N$, $f_n(z)$ has a zero in $|z - z_0| < \delta < \varepsilon$.

Poisson integral formula (Part 46)

Let $h(z)$ be a harmonic function, i.e., the real part of a function $f(z)$ regular in a domain $|z| < 1 + \varepsilon$, with $\varepsilon > 0$. Then, for $r < 1$,

$$h(re^{i\theta}) = \frac{1}{2\pi} \int_{-\pi}^{\pi} \frac{1 - r^2}{1 - 2r \cos(\theta - \varphi) + r^2} h(e^{i\varphi}) d\varphi$$

There is also a similar formula for the imaginary part of $f(z)$.
Proof. We first consider the special case when

$$f(z) = \sum_{n=0}^{\infty} a_n z^n , \qquad a_n = \text{real}$$

In this case, if we set

$$f(z) = f(re^{i\theta}) = h(z) + ik(z)$$

then

(i) $$f(re^{-i\theta}) = h(z) - ik(z)$$

For the sake of simplicity, we denote the values of $h(z)$, $k(z)$ at the points $z = re^{i\theta}$, $\xi = e^{i\varphi}$ by

$$h, k; h_1, k_1,$$

respectively. Then, by Cauchy,s integral formula, we obtain

(ii) $$h + ik = \frac{1}{2\pi i} \int \frac{h_1 + ik_1}{\xi - z} d\xi$$

$$= \frac{1}{2\pi} \int_0^{2\pi} \frac{(h_1 + ik_1)e^{i\varphi}}{e^{i\varphi} - re^{i\theta}} d\varphi$$

On the other hand, since the point $1/z$ is outside of the domain $|\xi| \leqq 1$, Cauchy's integral theorem implies

$$0 = \frac{1}{2\pi i} \int \frac{h_1 + ik_1}{\xi - 1/z} d\xi = \frac{1}{2\pi} \int_0^{2\pi} \frac{(h_1 + ik_1)e^{i\varphi}}{e^{i\varphi} - r^{-1}e^{-i\theta}} d\varphi$$

Hence, because of (i), we obtain

$$\frac{1}{2\pi} \int_0^{2\pi} \frac{(h_1 - ik_1)e^{-i\varphi}}{e^{-i\varphi} - r^{-1}e^{-i\theta}} d\varphi = 0$$

so that

(iii)
$$\frac{-1}{2\pi} \int_0^{2\pi} \frac{(h_1 - ik_1)re^{i\theta}}{re^{i\theta} - e^{i\varphi}} d\varphi = 0$$

Therefore, adding (ii) and (iii), we obtain

$$h + ik = \frac{1}{2\pi} \int_0^{2\pi} \left\{ h_1 \frac{e^{i\varphi} + re^{i\theta}}{e^{i\varphi} - re^{i\theta}} + ik_1 \right\} d\varphi$$

Then, taking the real parts of the both sides, we obtain the Poisson integral formula.

In the general case when (i) is not satisfied, that is,

$$f(z) = \sum_{n=0}^{\infty} a_n z^n , \qquad a_n = \alpha_n + i\beta_n$$

we have

$$f(z) = \sum_{n=0}^{\infty} \alpha_n z^n + i \sum_{n=0}^{\infty} \beta_n z^n$$
$$= f_1(z) + if_2(z)$$

Then, both $f_1(z)$ and $f_2(z)$ satisfy the condition (i). Hence, we can obtain the Poisson formula for $\mathscr{R}[f_1(z)]$, $\mathscr{I}[f_2(z)]$, then, by writing

$$\mathscr{R}[f(z)] = \mathscr{R}[f_1(z)] - \mathscr{I}[f_2(z)]$$

we can obtain the Poisson integral formula for $\mathscr{R}[f(z)]$.

BIBLIOGRAPHY

The bibliography contains books and papers related directly to this book, and it is not to be considered complete.

Concerning Chapter 1, see

1. Bieberbach, L., *Theorie der Differentialgleichungen*, Berlin, 3te Auflage, 1930.
2. Goursat, É., *Cours d'Analyse*, II et III, Paris, 1923.
3. Hadamard, J., *Cours d'Analyse*, II, Paris, 1930.
4. Ince, E. L., *Ordinary Differential Equations*, London, 1927.
5. Vallée Poussin, de la, *Cours d'Analyse*, II, Paris, 1925.

For the further study of differential equations, see

6. Coddington, E. A. and Levinson, N., *Theory of Ordinary Differential Equations*, New York, 1955.
7. Picard, É., *Traité d'Analyse*, II et III, Paris, 1928.

The boundary value problem for linear ordinary differential equations of the second order (Chapter 2) is also discussed in the framework of the theory of differential equations without using the theory of integral equations, see 1., 4. An especially interesting method due to Prüfer is found in 1.

Concerning Chapter 2 and 4, see 2.,

8. Bôcher, M., *An Introduction to the Study of Integral Equations*, Cambridge, 1909.
9. Courant, R. and Hilbert, D., *Methods of Mathematical Physics* 1, New York, 1953.
10. Hellinger, E. and Toeplitz, O., *Integralgleichungen und Gleichungen mit unendlich vielen Unbekannten*, Leipzig und Berlin, 1928.
11. Hilbert, D., *Grundzüge einer allgemeinen Theorie der linearen Integralgleichungen*, Leipzig und Berlin, 1912.
12. Kneser, A., *Die Integralgleichungen und ihre Anwendungen in der mathematischen Physik*, Braunshweig, 2te Auflage, 1922.
13. Lalésco, T., *Introduction à la théorie des équations intégrales*, Paris, 1912.
14. Vivanti, G. and Schwank, F., *Elemente der Theorie der linearen Integralgleichungen*, Hannover, 1929.
15. Volterra, V., *Leçons sur les équations intégrales et les équations intégro-différentielles*, Paris, 1913.

The complete bibliography, up to 1928, is found in 10. The existence of eigenvalues is proved in various ways. The method in Part 20 is one which is used in the theory of Hilbert space and is different from those in the books mentioned above, except for 9.

The proof of Fredholm's alternative theorem (§ 1, Chapter 3) is due to E. Schmidt:

16. Schmidt, E., "Zur Theorie der linearen und nicht linearen Integralgleichungen," *Math. Ann.*, **64**, 1907, pp. 161–174.

Usually, the theorem is proved by making use of the so-called Fredholm determinant.

The general expansion theorem (Chapter 5) goes back to:

17. Weyl, H., "Über gewöhnliche lineare Differentialgleichungen mit Singularitäten und die zugehörigen Entwicklungen willkürlicher Funktionen," *Math. Ann.*, **68**, 1910, pp. 220–269, which was developed further by M. H. Stone on the basis of the Hilbert space theory:

18. Stone, M. H., *Linear Transformations in Hilbert Space*, New York, 1932.

The Weyl-Stone theory was completed by E. C. Titchmarsh and K. Kodaira, by giving the formula to calculate the density matrix:

19. Kodaira, K., "The eigenvalue problem for ordinary differential equations of the second order and Heisenberg's theory of S-matrices," *Amer. J. Math.*, **71**, 1949, pp. 921–945.

20. Titchmarsh, E. C., *Eigenfunction Expansions Associated with Second-Order Differential Equations*, Oxford, 1946.

The elementary proof in Chapter 5 is due to the author. (Yosida, K., "On Titchmarsh-Kodaira's formula concerning Weyl-Stone's eigenfunction expansion," *Nagoya Math. J.*, **1**, 1950, pp. 49–58; the correction to the preceding paper, *Nagoya Math. J.*, **6**, 1953, pp. 187–188.)

Another elementary proof was also independently given by N. Levinson:

21 Levinson, N., "A simplified proof of the expansion theorem for singular second order linear differential equations," *Duke Math. J.*, **18**, 1951, pp. 57–71.

Concerning Chapter 6, see 13.,

22. Lichtenstein, L., *Vorlesungen über einige Klassen nichtlinearer Integralgleichungen und Integro-Differentialgleichungen*, Berlin, 1931 and Kapitel VI in:

23. Ljusternik, L. A. and Soboleff, W. I., *Elemente der Funktionalanalysis*, Berlin, 1955.

INDEX

A CATALOG OF SELECTED

DOVER BOOKS
IN SCIENCE AND MATHEMATICS

DOVER BOOKS
IN SCIENCE AND MATHEMATICS

QUALITATIVE THEORY OF DIFFERENTIAL EQUATIONS, V.V. Nemytskii and V.V. Stepanov. Classic graduate-level text by two prominent Soviet mathematicians covers classical differential equations as well as topological dynamics and erqodic theory. Bibliographies. 523pp. 5⅜ × 8½. 65954-2 Pa. $10.95

MATRICES AND LINEAR ALGEBRA, Hans Schneider and George Phillip Barker. Basic textbook covers theory of matrices and its applications to systems of linear equations and related topics such as determinants, eigenvalues and differential equations. Numerous exercises. 432pp. 5⅜ × 8½. 66014-1 Pa. $8.95

QUANTUM THEORY, David Bohm. This advanced undergraduate-level text presents the quantum theory in terms of qualitative and imaginative concepts, followed by specific applications worked out in mathematical detail. Preface. Index. 655pp. 5⅜ × 8½. 65969-0 Pa. $10.95

ATOMIC PHYSICS (8th edition), Max Born. Nobel laureate's lucid treatment of kinetic theory of gases, elementary particles, nuclear atom, wave-corpuscles, atomic structure and spectral lines, much more. Over 40 appendices, bibliography. 495pp. 5⅜ × 8½. 65984-4 Pa. $11.95

ELECTRONIC STRUCTURE AND THE PROPERTIES OF SOLIDS: The Physics of the Chemical Bond, Walter A. Harrison. Innovative text offers basic understanding of the electronic structure of covalent and ionic solids, simple metals, transition metals and their compounds. Problems. 1980 edition. 582pp. 6⅛ × 9¼. 66021-4 Pa. $14.95

BOUNDARY VALUE PROBLEMS OF HEAT CONDUCTION, M. Necati Özisik. Systematic, comprehensive treatment of modern mathematical methods of solving problems in heat conduction and diffusion. Numerous examples and problems. Selected references. Appendices. 505pp. 5⅜ × 8½. 65990-9 Pa. $11.95

A SHORT HISTORY OF CHEMISTRY (3rd edition), J.R. Partington. Classic exposition explores origins of chemistry, alchemy, early medical chemistry, nature of atmosphere, theory of valency, laws and structure of atomic theory, much more. 428pp. 5⅜ × 8½. (Available in U.S. only) 65977-1 Pa. $10.95

A HISTORY OF ASTRONOMY, A. Pannekoek. Well-balanced, carefully reasoned study covers such topics as Ptolemaic theory, work of Copernicus, Kepler, Newton, Eddington's work on stars, much more. Illustrated. References. 521pp. 5⅜ × 8½. 65994-1 Pa. $11.95

PRINCIPLES OF METEOROLOGICAL ANALYSIS, Walter J. Saucier. Highly respected, abundantly illustrated classic reviews atmospheric variables, hydrostatics, static stability, various analyses (scalar, cross-section, isobaric, isentropic, more). For intermediate meteorology students. 454pp. 6⅛ × 9¼. 65979-8 Pa. $12.95

RELATIVITY, THERMODYNAMICS AND COSMOLOGY, Richard C. Tolman. Landmark study extends thermodynamics to special, general relativity; also applications of relativistic mechanics, thermodynamics to cosmological models. 501pp. 5⅜ × 8½. 65383-8 Pa. $11.95

APPLIED ANALYSIS, Cornelius Lanczos. Classic work on analysis and design of finite processes for approximating solution of analytical problems. Algebraic equations, matrices, harmonic analysis, quadrature methods, much more. 559pp. 5⅜ × 8½. 65656-X Pa. $11.95

SPECIAL RELATIVITY FOR PHYSICISTS, G. Stephenson and C.W. Kilmister. Concise elegant account for nonspecialists. Lorentz transformation, optical and dynamical applications, more. Bibliography. 108pp. 5⅜ × 8½. 65519-9 Pa. $3.95

INTRODUCTION TO ANALYSIS, Maxwell Rosenlicht. Unusually clear, accessible coverage of set theory, real number system, metric spaces, continuous functions, Riemann integration, multiple integrals, more. Wide range of problems. Undergraduate level. Bibliography. 254pp. 5⅜ × 8½. 65038-3 Pa. $7.00

INTRODUCTION TO QUANTUM MECHANICS With Applications to Chemistry, Linus Pauling & E. Bright Wilson, Jr. Classic undergraduate text by Nobel Prize winner applies quantum mechanics to chemical and physical problems. Numerous tables and figures enhance the text. Chapter bibliographies. Appendices. Index. 468pp. 5⅜ × 8½. 64871-0 Pa. $9.95

ASYMPTOTIC EXPANSIONS OF INTEGRALS, Norman Bleistein & Richard A. Handelsman. Best introduction to important field with applications in a variety of scientific disciplines. New preface. Problems. Diagrams. Tables. Bibliography. Index. 448pp. 5⅜ × 8½. 65082-0 Pa. $10.95

MATHEMATICS APPLIED TO CONTINUUM MECHANICS, Lee A. Segel. Analyzes models of fluid flow and solid deformation. For upper-level math, science and engineering students. 608pp. 5⅜ × 8½. 65369-2 Pa. $12.95

ELEMENTS OF REAL ANALYSIS, David A. Sprecher. Classic text covers fundamental concepts, real number system, point sets, functions of a real variable, Fourier series, much more. Over 500 exercises. 352pp. 5⅜ × 8½. 65385-4 Pa. $8.95

PHYSICAL PRINCIPLES OF THE QUANTUM THEORY, Werner Heisenberg. Nobel Laureate discusses quantum theory, uncertainty, wave mechanics, work of Dirac, Schroedinger, Compton, Wilson, Einstein, etc. 184pp. 5⅜ × 8½. 60113-7 Pa. $4.95

INTRODUCTORY REAL ANALYSIS, A.N. Kolmogorov, S.V. Fomin. Translated by Richard A. Silverman. Self-contained, evenly paced introduction to real and functional analysis. Some 350 problems. 403pp. 5⅜ × 8½. 61226-0 Pa. $7.95

PROBLEMS AND SOLUTIONS IN QUANTUM CHEMISTRY AND PHYSICS, Charles S. Johnson, Jr. and Lee G. Pedersen. Unusually varied problems, detailed solutions in coverage of quantum mechanics, wave mechanics, angular momentum, molecular spectroscopy, scattering theory, more. 280 problems plus 139 supplementary exercises. 430pp. 6½ × 9¼. 65236-X Pa. $10.95

ASYMPTOTIC METHODS IN ANALYSIS, N.G. de Bruijn. An inexpensive, comprehensive guide to asymptotic methods—the pioneering work that teaches by explaining worked examples in detail. Index. 224pp. 5⅜ × 8½. 64221-6 Pa. $5.95

OPTICAL RESONANCE AND TWO-LEVEL ATOMS, L. Allen and J.H. Eberly. Clear, comprehensive introduction to basic principles behind all quantum optical resonance phenomena. 53 illustrations. Preface. Index. 256pp. 5⅜ × 8½. 65533-4 Pa. $6.95

COMPLEX VARIABLES, Francis J. Flanigan. Unusual approach, delaying complex algebra till harmonic functions have been analyzed from real variable viewpoint. Includes problems with answers. 364pp. 5⅜ × 8½. 61388-7 Pa. $7.95

ATOMIC SPECTRA AND ATOMIC STRUCTURE, Gerhard Herzberg. One of best introductions; especially for specialist in other fields. Treatment is physical rather than mathematical. 80 illustrations. 257pp. 5⅜ × 8½. 60115-3 Pa. $4.95

APPLIED COMPLEX VARIABLES, John W. Dettman. Step-by-step coverage of fundamentals of analytic function theory—plus lucid exposition of 5 important applications: Potential Theory; Ordinary Differential Equations; Fourier Transforms; Laplace Transforms; Asymptotic Expansions. 66 figures. Exercises at chapter ends. 512pp. 5⅜ × 8½. 64670-X Pa. $10.95

ULTRASONIC ABSORPTION: An Introduction to the Theory of Sound Absorption and Dispersion in Gases, Liquids and Solids, A.B. Bhatia. Standard reference in the field provides a clear, systematically organized introductory review of fundamental concepts for advanced graduate students, research workers. Numerous diagrams. Bibliography. 440pp. 5⅜ × 8½. 64917-2 Pa. $8.95

UNBOUNDED LINEAR OPERATORS: Theory and Applications, Seymour Goldberg. Classic presents systematic treatment of the theory of unbounded linear operators in normed linear spaces with applications to differential equations. Bibliography. 199pp. 5⅜ × 8½. 64830-3 Pa. $7.00

LIGHT SCATTERING BY SMALL PARTICLES, H.C. van de Hulst. Comprehensive treatment including full range of useful approximation methods for researchers in chemistry, meteorology and astronomy. 44 illustrations. 470pp. 5⅜ × 8½. 64228-3 Pa. $9.95

CONFORMAL MAPPING ON RIEMANN SURFACES, Harvey Cohn. Lucid, insightful book presents ideal coverage of subject. 334 exercises make book perfect for self-study. 55 figures. 352pp. 5⅜ × 8¼. 64025-6 Pa. $8.95

OPTICKS, Sir Isaac Newton. Newton's own experiments with spectroscopy, colors, lenses, reflection, refraction, etc., in language the layman can follow. Foreword by Albert Einstein. 532pp. 5⅜ × 8½. 60205-2 Pa. $8.95

GENERALIZED INTEGRAL TRANSFORMATIONS, A.H. Zemanian. Graduate-level study of recent generalizations of the Laplace, Mellin, Hankel, K. Weierstrass, convolution and other simple transformations. Bibliography. 320pp. 5⅜ × 8½. 65375-7 Pa. $7.95

ORDINARY DIFFERENTIAL EQUATIONS, Morris Tenenbaum and Harry Pollard. Exhaustive survey of ordinary differential equations for undergraduates in mathematics, engineering, science. Thorough analysis of theorems. Diagrams. Bibliography. Index. 818pp. 5⅜ × 8½. 64940-7 Pa. $15.95

STATISTICAL MECHANICS: Principles and Applications, Terrell L. Hill. Standard text covers fundamentals of statistical mechanics, applications to fluctuation theory, imperfect gases, distribution functions, more. 448pp. 5⅜ × 8½. 65390-0 Pa. $9.95

ORDINARY DIFFERENTIAL EQUATIONS AND STABILITY THEORY: An Introduction, David A. Sánchez. Brief, modern treatment. Linear equation, stability theory for autonomous and nonautonomous systems, etc. 164pp. 5⅜ × 8¼. 63828-6 Pa. $4.95

THIRTY YEARS THAT SHOOK PHYSICS: The Story of Quantum Theory, George Gamow. Lucid, accessible introduction to influential theory of energy and matter. Careful explanations of Dirac's anti-particles, Bohr's model of the atom, much more. 12 plates. Numerous drawings. 240pp. 5⅜ × 8½. 24895-X Pa. $5.95

ORDINARY DIFFERENTIAL EQUATIONS, I.G. Petrovski. Covers basic concepts, some differential equations and such aspects of the general theory as Euler lines, Arzel's theorem, Peano's existence theorem, Osgood's uniqueness theorem, more. 45 figures. Problems. Bibliography. Index. xi + 232pp. 5⅜ × 8½. 64683-1 Pa. $6.00

GREAT EXPERIMENTS IN PHYSICS: Firsthand Accounts from Galileo to Einstein, edited by Morris H. Shamos. 25 crucial discoveries: Newton's laws of motion, Chadwick's study of the neutron, Hertz on electromagnetic waves, more. Original accounts clearly annotated. 370pp. 5⅜ × 8½. 25346-5 Pa. $8.95

INTRODUCTION TO PARTIAL DIFFERENTIAL EQUATIONS WITH APPLICATIONS, E.C. Zachmanoglou and Dale W. Thoe. Essentials of partial differential equations applied to common problems in engineering and the physical sciences. Problems and answers. 416pp. 5⅜ × 8½. 65251-3 Pa. $9.95

BURNHAM'S CELESTIAL HANDBOOK, Robert Burnham, Jr. Thorough guide to the stars beyond our solar system. Exhaustive treatment. Alphabetical by constellation: Andromeda to Cetus in Vol. 1; Chamaeleon to Orion in Vol. 2; and Pavo to Vulpecula in Vol. 3. Hundreds of illustrations. Index in Vol. 3. 2,000pp. 6⅛ × 9¼. 23567-X, 23568-8, 23673-0 Pa., Three-vol. set $38.85

ASYMPTOTIC EXPANSIONS FOR ORDINARY DIFFERENTIAL EQUATIONS, Wolfgang Wasow. Outstanding text covers asymptotic power series, Jordan's canonical form, turning point problems, singular perturbations, much more. Problems. 384pp. 5⅜ × 8½. 65456-7 Pa. $8.95

AMATEUR ASTRONOMER'S HANDBOOK, J.B. Sidgwick. Timeless, comprehensive coverage of telescopes, mirrors, lenses, mountings, telescope drives, micrometers, spectroscopes, more. 189 illustrations. 576pp. 5⅜ × 8¼. 24034-7 Pa. $8.95

HANDBOOK OF MATHEMATICAL FUNCTIONS WITH FORMULAS, GRAPHS, AND MATHEMATICAL TABLES, edited by Milton Abramowitz and Irene A. Stegun. Vast compendium: 29 sets of tables, some to as high as 20 places. 1,046pp. 8 × 10½. 61272-4 Pa. $21.95

MATHEMATICAL METHODS IN PHYSICS AND ENGINEERING, John W. Dettman. Algebraically based approach to vectors, mapping, diffraction, other topics in applied math. Also generalized functions, analytic function theory, more. Exercises. 448pp. 5⅜ × 8¼. 65649-7 Pa. $8.95

A SURVEY OF NUMERICAL MATHEMATICS, David M. Young and Robert Todd Gregory. Broad self-contained coverage of computer-oriented numerical algorithms for solving various types of mathematical problems in linear algebra, ordinary and partial, differential equations, much more. Exercises. Total of 1,248pp. 5⅜ × 8½. Two volumes. Vol. I 65691-8 Pa. $13.95
Vol. II 65692-6 Pa. $13.95

TENSOR ANALYSIS FOR PHYSICISTS, J.A. Schouten. Concise exposition of the mathematical basis of tensor analysis, integrated with well-chosen physical examples of the theory. Exercises. Index. Bibliography. 289pp. 5⅜ × 8½. 65582-2 Pa. $7.95

INTRODUCTION TO NUMERICAL ANALYSIS (2nd Edition), F.B. Hildebrand. Classic, fundamental treatment covers computation, approximation, interpolation, numerical differentiation and integration, other topics. 150 new problems. 669pp. 5⅜ × 8½. 65363-3 Pa. $13.95

INVESTIGATIONS ON THE THEORY OF THE BROWNIAN MOVEMENT, Albert Einstein. Five papers (1905–8) investigating dynamics of Brownian motion and evolving elementary theory. Notes by R. Fürth. 122pp. 5⅜ × 8½. 60304-0 Pa. $3.95

NUMERICAL METHODS FOR SCIENTISTS AND ENGINEERS, Richard Hamming. Classic text stresses frequency approach in coverage of algorithms, polynomial approximation, Fourier approximation, exponential approximation, other topics. Revised and enlarged 2nd edition. 721pp. 5⅜ × 8½. 65241-6 Pa. $14.95

AN INTRODUCTION TO STATISTICAL THERMODYNAMICS, Terrell L. Hill. Excellent basic text offers wide-ranging coverage of quantum statistical mechanics, systems of interacting molecules, quantum statistics, more. 523pp. 5⅜ × 8½. 65242-4 Pa. $10.95

ELEMENTARY DIFFERENTIAL EQUATIONS, William Ted Martin and Eric Reissner. Exceptionally, clear comprehensive introduction at undergraduate level. Nature and origin of differential equations, differential equations of first, second and higher orders. Picard's Theorem, much more. Problems with solutions. 331pp. 5⅜ × 8½. 65024-3 Pa. $8.95

STATISTICAL PHYSICS, Gregory H. Wannier. Classic text combines thermodynamics, statistical mechanics and kinetic theory in one unified presentation of thermal physics. Problems with solutions. Bibliography. 532pp. 5⅜ × 8½. 65401-X Pa. $10.95

NUMERICAL METHODS FOR SCIENTISTS AND ENGINEERS, Richard Hamming. Classic text stresses frequency approach in coverage of algorithms, polynomial approximation, Fourier approximation, exponential approximation, other topics. Revised and enlarged 2nd edition. 721pp. 5⅜ × 8½.
65241-6 Pa. $14.95

THEORETICAL SOLID STATE PHYSICS, Vol. I: Perfect Lattices in Equilibrium; Vol. II: Non-Equilibrium and Disorder, William Jones and Norman H. March. Monumental reference work covers fundamental theory of equilibrium properties of perfect crystalline solids, non-equilibrium properties, defects and disordered systems. Appendices. Problems. Preface. Diagrams. Index. Bibliography. Total of 1,301pp. 5⅜ × 8½. Two volumes.
Vol. I 65015-4 Pa. $12.95
Vol. II 65016-2 Pa. $12.95

OPTIMIZATION THEORY WITH APPLICATIONS, Donald A. Pierre. Broadspectrum approach to important topic. Classical theory of minima and maxima, calculus of variations, simplex technique and linear programming, more. Many problems, examples. 640pp. 5⅜ × 8½.
65205-X Pa. $12.95

THE MODERN THEORY OF SOLIDS, Frederick Seitz. First inexpensive edition of classic work on theory of ionic crystals, free-electron theory of metals and semiconductors, molecular binding, much more. 736pp. 5⅜ × 8½.
65482-6 Pa. $14.95

ESSAYS ON THE THEORY OF NUMBERS, Richard Dedekind. Two classic essays by great German mathematician: on the theory of irrational numbers; and on transfinite numbers and properties of natural numbers. 115pp. 5⅜ × 8½.
21010-3 Pa. $4.95

THE FUNCTIONS OF MATHEMATICAL PHYSICS, Harry Hochstadt. Comprehensive treatment of orthogonal polynomials, hypergeometric functions, Hill's equation, much more. Bibliography. Index. 322pp. 5⅜ × 8½.   65214-9 Pa. $8.95

NUMBER THEORY AND ITS HISTORY, Oystein Ore. Unusually clear, accessible introduction covers counting, properties of numbers, prime numbers, much more. Bibliography. 380pp. 5⅜ × 8½.   65620-9 Pa. $8.95

THE VARIATIONAL PRINCIPLES OF MECHANICS, Cornelius Lanczos. Graduate level coverage of calculus of variations, equations of motion, relativistic mechanics, more. First inexpensive paperbound edition of classic treatise. Index. Bibliography. 418pp. 5⅜ × 8½.   65067-7 Pa. $10.95

MATHEMATICAL TABLES AND FORMULAS, Robert D. Carmichael and Edwin R. Smith. Logarithms, sines, tangents, trig functions, powers, roots, reciprocals, exponential and hyperbolic functions, formulas and theorems. 269pp. 5⅜ × 8½.   60111-0 Pa. $5.95

THEORETICAL PHYSICS, Georg Joos, with Ira M. Freeman. Classic overview covers essential math, mechanics, electromagnetic theory, thermodynamics, quantum mechanics, nuclear physics, other topics. First paperback edition. xxiii + 885pp. 5⅜ × 8½.   65227-0 Pa. $17.95

THE ELECTROMAGNETIC FIELD, Albert Shadowitz. Comprehensive undergraduate text covers basics of electric and magnetic fields, builds up to electromagnetic theory. Also related topics, including relativity. Over 900 problems. 768pp. 5⅜ × 8¼. 65660-8 Pa. $15.95

FOURIER SERIES, Georgi P. Tolstov. Translated by Richard A. Silverman. A valuable addition to the literature on the subject, moving clearly from subject to subject and theorem to theorem. 107 problems, answers. 336pp. 5⅜ × 8½. 63317-9 Pa. $7.95

THEORY OF ELECTROMAGNETIC WAVE PROPAGATION, Charles Herach Papas. Graduate-level study discusses the Maxwell field equations, radiation from wire antennas, the Doppler effect and more. xiii + 244pp. 5⅜ × 8½. 65678-0 Pa. $6.95

DISTRIBUTION THEORY AND TRANSFORM ANALYSIS: An Introduction to Generalized Functions, with Applications, A.H. Zemanian. Provides basics of distribution theory, describes generalized Fourier and Laplace transformations. Numerous problems. 384pp. 5⅜ × 8½. 65479-6 Pa. $8.95

THE PHYSICS OF WAVES, William C. Elmore and Mark A. Heald. Unique overview of classical wave theory. Acoustics, optics, electromagnetic radiation, more. Ideal as classroom text or for self-study. Problems. 477pp. 5⅜ × 8½. 64926-1 Pa. $10.95

CALCULUS OF VARIATIONS WITH APPLICATIONS, George M. Ewing. Applications-oriented introduction to variational theory develops insight and promotes understanding of specialized books, research papers. Suitable for advanced undergraduate/graduate students as primary, supplementary text. 352pp. 5⅜ × 8½. 64856-7 Pa. $8.50

A TREATISE ON ELECTRICITY AND MAGNETISM, James Clerk Maxwell. Important foundation work of modern physics. Brings to final form Maxwell's theory of electromagnetism and rigorously derives his general equations of field theory. 1,084pp. 5⅜ × 8½. 60636-8, 60637-6 Pa., Two-vol. set $19.00

AN INTRODUCTION TO THE CALCULUS OF VARIATIONS, Charles Fox. Graduate-level text covers variations of an integral, isoperimetrical problems, least action, special relativity, approximations, more. References. 279pp. 5⅜ × 8½. 65499-0 Pa. $6.95

HYDRODYNAMIC AND HYDROMAGNETIC STABILITY, S. Chandrasekhar. Lucid examination of the Rayleigh-Benard problem; clear coverage of the theory of instabilities causing convection. 704pp. 5⅜ × 8¼. 64071-X Pa. $12.95

CALCULUS OF VARIATIONS, Robert Weinstock. Basic introduction covering isoperimetric problems, theory of elasticity, quantum mechanics, electrostatics, etc. Exercises throughout. 326pp. 5⅜ × 8½. 63069-2 Pa. $7.95

DYNAMICS OF FLUIDS IN POROUS MEDIA, Jacob Bear. For advanced students of ground water hydrology, soil mechanics and physics, drainage and irrigation engineering and more. 335 illustrations. Exercises, with answers. 784pp. 6⅛ × 9¼. 65675-6 Pa. $19.95

CHALLENGING MATHEMATICAL PROBLEMS WITH ELEMENTARY SOLUTIONS, A.M. Yaglom and I.M. Yaglom. Over 170 challenging problems on probability theory, combinatorial analysis, points and lines, topology, convex polygons, many other topics. Solutions. Total of 445pp. 5⅜ × 8½. Two-vol. set.

Vol. I 65536-9 Pa. $5.95
Vol. II 65537-7 Pa. $5.95

FIFTY CHALLENGING PROBLEMS IN PROBABILITY WITH SOLUTIONS, Frederick Mosteller. Remarkable puzzlers, graded in difficulty, illustrate elementary and advanced aspects of probability. Detailed solutions. 88pp. 5⅜ × 8½.
65355-2 Pa. $3.95

EXPERIMENTS IN TOPOLOGY, Stephen Barr. Classic, lively explanation of one of the byways of mathematics. Klein bottles, Moebius strips, projective planes, map coloring, problem of the Koenigsberg bridges, much more, described with clarity and wit. 43 figures. 210pp. 5⅜ × 8½.
25933-1 Pa. $4.95

RELATIVITY IN ILLUSTRATIONS, Jacob T. Schwartz. Clear non-technical treatment makes relativity more accessible than ever before. Over 60 drawings illustrate concepts more clearly than text alone. Only high school geometry needed. Bibliography. 128pp. 6⅛ × 9¼.
25965-X Pa. $5.95

AN INTRODUCTION TO ORDINARY DIFFERENTIAL EQUATIONS, Earl A. Coddington. A thorough and systematic first course in elementary differential equations for undergraduates in mathematics and science, with many exercises and problems (with answers). Index. 304pp. 5⅜ × 8¼.
65942-9 Pa. $7.95

FOURIER SERIES AND ORTHOGONAL FUNCTIONS, Harry F. Davis. An incisive text combining theory and practical example to introduce Fourier series, orthogonal functions and applications of the Fourier method to boundary-value problems. 570 exercises. Answers and notes. 416pp. 5⅜ × 8½.
65973-9 Pa. $8.95

THE THOERY OF BRANCHING PROCESSES, Theodore E. Harris. First systematic, comprehensive treatment of branching (i.e. multiplicative) processes and their applications. Galton-Watson model, Markov branching processes, electron-photon cascade, many other topics. Rigorous proofs. Bibliography. 240pp. 5⅜ × 8½.
65952-6 Pa. $6.95

AN INTRODUCTION TO ALGEBRAIC STRUCTURES, Joseph Landin. Superb self-contained text covers "abstract algebra": sets and numbers, theory of groups, theory of rings, much more. Numerous well-chosen examples, exercises. 247pp. 5⅜ × 8½.
65940-2 Pa. $6.95

GAMES AND DECISIONS: Introduction and Critical Survey, R. Duncan Luce and Howard Raiffa. Superb non-technical introduction to game theory, primarily applied to social sciences. Utility theory, zero-sum games, n-person games, decision-making, much more. Bibliography. 509pp. 5⅜ × 8½. 65943-7 Pa. $10.95

Prices subject to change without notice.
Available at your book dealer or write for free Mathematics and Science Catalog to Dept. GI, Dover Publications, Inc., 31 East 2nd St., Mineola, N.Y. 11501. Dover publishes more than 175 books each year on science, elementary and advanced mathematics, biology, music, art, literary history, social sciences and other areas.